Peschke/von Olshausen   Cable Systems for High
and Extra-High Voltage

Cable Systems for High and Extra-High Voltage

Development, Manufacture, Testing, Installation and Operation of Cables and their Accessories

By Egon F. Peschke and Rainer von Olshausen

Publicis MCD Verlag

Die Deutsche Bibliothek – CIP-Einheitsaufnahme

Peschke, Egon F.:
Cable systems for high and extra high voltage : development,
manufacture, testing, installation and operation of cables and their
accessories / by Egon F. Peschke and Rainer von Olshausen. [Issued
by: Pirelli Kabel und Systeme GmbH, Berlin]. – Erlangen ; Munich :
Publicis-MCD-Verl., 1999
   Dt. Ausg. u.d.T.: Peschke, Egon F.: Kabelanlagen für Hoch- und
   Höchstspannung
   ISBN 3-89578-118-5

ISBN 3-89578-118-5

Issued by: Pirelli Kabel und Systeme GmbH, Berlin
Publisher: Publicis MCD Verlag, Erlangen and Munich

Printed in Germany

Preface

This book is based on two lectures on the subject of power cables given by the authors to students of power engineering at the Munich Technical University and the University of Hanover. As the two lectures were prepared completely independently of each other and have been further developed over a number of years, it has been a stimulating task to combine the material from these lectures and to publish it in a bound volume. The benefits of the experience the authors have gained from their many years work in senior positions in the Power Cables Division of Siemens AG, their many contacts and discussions with specialists within the company and other manufacturers, their active participation in specialist committees of Cigré, e.g. as Chairman of Study Committee No. 21 "High Voltage Insulated Cables", of VDE/ETG, etc., and, not least, the results of the intensive exchange of experience with material suppliers and operators of cable systems have been included in this work.

This book is published at a time when important developments in the high and extra high voltage cable sector are reaching their conclusion and their application in technical practice is imminent, and when the manufacturers' development centers and research laboratories are busy with completely new activities. These are, on the one hand, the replacement of the traditional oil-filled paper-insulated cable by polymer insulated cable using cross-linked polyethylene (XLPE) insulation for 420 to 525 kV and, on the other hand, the use of high-temperature superconductors (HTSC) in power cables for selected high-power systems. Both themes are examined in detail: polymer insulated cables with XLPE dielectric and their accessories as the major subject of the book and HTSC cable as an interesting vision of the future for the design of powerful transmission systems for the underground transmission of large amounts of energy. Also gas-insulated lines, whose application has up to now been limited to relatively short links at high operating voltages and high power, are considered as possible alternatives to overhead power transmission lines.

Naturally, all other types of high and very high voltage cable currently in use are discussed, together with the components necessary for a functional system. This includes the oil-filled paper-insulated cables already mentioned as well as the gas-pressure cables that are commonly used in the German 110 kV network.

In addition to the general properties, design, materials, manufacturing technology and quality assurance of the cables, particular attention is paid to development and dimensioning including the difficult problem of a reliable forecast of service life in the case of solid insulation materials. The many years of professional experience of the authors assures that the necessary theoretical basis of materials, process technology and physics is treated in a balanced relationship with the practical aspects of cable technology.

Based as it is on lectures at technical universities, this book will be of particular interest to students of electrical engineering who are specialising in power systems engineering. It will also be invaluable to technicians and engineers working for power supply companies or in the cable industry.

Berlin, October 1999   Egon F. Peschke, Rainer v. Olshausen

Foreword

The German version of this book was first published by Siemens in June 1998 and plans were made at that time to publish an English version during 1999. In October 1998, however, Pirelli Cavi e Sistemi S.p.A acquired the Power Cable Division of Siemens AG and consequently became involved in the decision making process concerning the publication and contents of the proposed English version.

The view taken by Pirelli is that this book represents a very important contribution to the dissemination of technical knowledge regarding cable systems for high and extra high voltage at a time when major publications in this field are few and far between. Its authors, Egon Peschke and Rainer von Olshausen, both very experienced and well known in the cable engineering community, have managed to write a clear, comprehensive and very readable text covering a wide spectrum of high and extra high voltage cable system solutions and applications.

As evidenced also by its wide ranging bibliography, this book, while reflecting the views of its authors and taking into strong consideration the lengthy and prestigious technical experience of Siemens, endeavours at the same time to cover objectively the most significant technologies associated with high and extra high voltage cables.

Pirelli has therefore decided to support the publication of a substantially unchanged English translation of the original German version in the certainty that this book will become an important and frequently used reference for cable system users, cable engineers and students of electrical engineering.

Milano, October 1999   Aldo Bolza

Contents

1 Introduction

1.1 Cables, Accessories and Cable Systems

Cable is a capital investment that has a long life and is of great economic importance. In 1995, the electrical supply industry investment reached around 14 thousand million DM in West Germany and the new federal states [1]. The investment in the electrical power supply was thus greater than in any other branch of industry. The proportion of this investment relating to transmission and distribution, i.e. for the network, is about 50% of this total. Half of this, i.e. a quarter of the total annual amount, is invested for renewal and expansion of the public power supply cable network. The world market for cable amounts to about 40 to 50 thousand million DM. About 2 thousand million DM of this is for high voltage and extra high voltage cables.

Underground transmission and distribution will become ever more important worldwide, i.e. the importance of cable and cable systems will increase in the medium and long term. A number of points support this prediction:

- The energy consumption of the major conurbations in industrialized countries will increase due to the increase in population and the demands of local public transport. In these locations, it will only be possible to meet this demand by using underground cables.

- The increasing environmental awareness of the public makes it difficult – especially in central Europe – to construct new overhead transmission systems, even outside the large urban areas, because of the associated indisputable "use of the countryside". The many demands for the substitution of overhead lines also increases the demand for cable.

- An additional requirement for cable results also from the fact that parts of the existing cable network are reaching the end of their life span and, in some cases, their technical and ecological condition makes replacement necessary. This applies particularly to cables with oil-filled paper insulation as these can suffer from oil losses due to leakage. As paper insulated cables have an expected life of 40 to 50 years, there will be an increasing need for re-investment in the next few years, particularly in the case of the 110 kV cables that were installed in the large cities up to the end of the 1950s.

- In many developing countries, there is considerable pent-up demand in respect of reliable power supply systems, especially in the big cities. For instance, a few years ago, an ambitious program of 150 kV cabling was implemented in several regions of Turkey.

- The present third world countries will, in the foreseeable future, gradually upgrade their power supply systems to meet the growing demand for electrical energy and here, too, this will be done predominantly using cables. An example of this is Bombay in India, with the expansion of their 110 kV and 220 kV cable network.

The demand for cables in all voltage ranges will increase further in the future, the environmental compatibility of this product being of decisive importance.

The cable itself forms just one part of the underground transmission system. A functional cable system always requires accessories in addition to the actual cable. These are the cable end terminations and, in the case of long cable runs, the connections between the individual lengths of cable delivered to site. Especially in the high voltage range, these accessories, which are available in a wide variety of forms, are the critical components of the cable system for a variety of reasons:

- The distribution of the electrical field which essentially differs locally from the ideal cylindrical form in the cable in spite of specific field control measures,

- The existence of boundary surfaces between the cable dielectric and the accessory insulation and

- The need to fit the accessories on site under conditions that are not comparable with those under which the cable was manufactured and even, depending on type, having to do this work by hand.

The basic requirement is that, in spite of all this, the accessories must be just as electrically, thermally and mechanically safe in operation as the cable itself. Special attention must therefore be paid to the accessories in their conception and development and during the design, installation, testing and, if necessary, maintenance of the cable system itself. Cable accessories are therefore dealt with in detail in their own chapter (Chapter 6) of this book.

A range of other components must be considered in high voltage cable systems. According to the type of construction and plant equipment, these include systems for the maintenance and monitoring of the pressure of the impregnation fluid or pressurizing gas and, e.g., for corrosion protection, the cable temperature along the cable run or at selected points, the penetration of moisture or the occurrence of partial discharges in polymer-insulated cables and their accessories. In particular, the last-mentioned monitoring systems have recently come to the fore in important transmission systems and are therefore treated under the general term "monitoring" in various places in this book.

1.2 Voltage Levels

A powerful three-phase network – consisting of overhead transmission lines, cables, transformers and switchgear – transports the energy from the power stations to the consumers. Because of the different functions, this network includes various voltage levels.

In Germany – with 4 levels– the extra high voltage network with its 380 kV and 220 kV cables provides long-distance transmission and power exchange with neighbouring countries. In conurbations, high-voltage networks with 110 kV cables additionally transmit energy from the power stations to the main points of consumption. The further distribution from here takes place via the 10 kV or 20 kV medium-voltage network. Most consumers are then connected to the low-voltage network.

Three-phase cables are now in use worldwide at all rated voltages between 400 V (generally known as "1 kV cable") and 500 kV (extra high voltage cable). Although the standards relating to safety (e.g. VDE 0100) already talk of high voltage for continuous potential differences of 250V between line and earth, the description "high voltage" is usually only applied to plant with a rated voltage between phases of more than 45 kV. The applicable standards also take this custom into account by formulating separate specifications for e.g. cables above 45 kV (e.g. DIN VDE 0276-632 to 0276-635, IEC 60840).

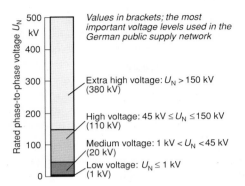

Figure 1.1
Voltage ranges of three-phase cables

The boundary between high and extra high voltage cables to be discussed in this book is based on a specification defined by VDEW:

- High voltage cable $45 \text{ kV} < U_N \leq 150 \text{ kV}$
- Extra high voltage cable $U_N > 150 \text{ kV}.$

The standardization of polymer high voltage cable is also oriented to this value; the specifications already mentioned, DIN VDE 0276-632 (formerly VDE 0263) and IEC 60840 only apply up to rated voltages of 150 kV. For clarity, Figure 1.1 shows again the various customary voltage ranges now in international use and the four most important voltage levels used in German networks.

It should be noted that, in this book, the rated voltage U_N is used exclusively for the designation of a voltage level or item of plant, in this case the cable (e.g. 110 kV cable). Conversely, the proof value for the cable insulation is the "maximum plant voltage" U_m. U_m is at least equal to the highest continuous permissible voltage and is 5 to 20% above the associated rated voltage.

1.3 Cables and Overhead Lines

Lines, i.e. overhead lines and cables, are important components of electrical networks. Their technical purpose, the reliable and economic transmission and distribution of electrical energy, can be carried out equally well by both overhead lines and cables. However, because of the different physical configuration of cables and overhead lines, there are various differences in their technical properties and operational behavior that must be taken into account in conjunction with other components of the network [2].

1.3.1 Technical and Operational Properties

Some of the decisive technical and operational properties of overhead lines and cables are compared below:

- *Insulation spacing and right-of-way requirements*
  It is relatively easy to deal with high voltages using overhead lines. The insulation material is atmospheric air. The problem is solved by sufficient spacing between the conductors and sufficiently long insulators. In the case of cables, solid or mixed

(solid-liquid) dielectrics with high dielectric strength are used, permitting relatively thin layers of insulation. The higher the operating voltage, the more difficult the design of the insulation becomes. The characteristic of high voltage cables is that high potential differences can be withstood at low spacings. As a result of this, the dielectric is stressed not only electrically but also thermally and, to some extent, mechanically. For instance, the spacing between conductor and ground for a 380 kV overhead line is 3000 mm, while a cable for the same rated voltage has an insulation thickness of only 25 to 27 mm. The ratio of the stress is therefore more than 100:1.

The small dimensions of the cable resulting from the insulation thickness is one of the reasons for the comparatively narrow rights of way required for cable systems. The width of the right of way required for the two systems is considerably different, the cable requiring much less space than the overhead line.

- *Operating capacitance and charging current*
  Because of their low insulation thicknesses compared with overhead lines, three-phase cables have a much greater operating capacitance per unit length. For voltage levels of 110 kV to 380 kV, the ratio is between 15:1 and 50:1. The high capacitance of the cable gives rise to a greater capacitive phase current (charging current), even at no load, and this has a limiting effect on the length of the power transmission cable. The charging current uses up more and more of the permissible current loading with increasing length of the cable until, finally, the cable is already thermally loaded to its limit without transmitting any power to the consumer. This means that in practice, without compensation, there is a definite limit to the maximum length, especially in the case of oil insulated cables [3] (see Section, 2.2.1, Figure 2.1 for details).

- *Transmitted power*
  In the case of overhead lines, the permissible temperature is determined by the mechanical strength of the conductor. The heat losses are easily dissipated in the surrounding air, so that high loads and periodical overloads are possible. Overhead lines are therefore usually not designed according to thermal considerations but according to their "Natural Load" P_{nat}. The thermal load, i.e. the transmission capacity up to the permissible thermal limit of the conductor, is a multiple of this value (approx. ratio 3:1 to 7:1). Overhead lines therefore provide a large overload reserve.

Conversely, in the case of cable systems, the transmission capacity is determined by the permissible thermal limit of the dielectric (max. permissible conductor temperature), the high thermal resistance of the parts of the cable and the soil conditions. Additional difficulties arise because the thermal resistance of the soil depends considerably on the moisture content and is therefore not a constant value and because additional losses (dielectric losses, sheath losses) have to be taken into account in the case of cables. As a result of this, the transmission capacity of high voltage cable systems has to be calculated very carefully from the thermal point of view. The thermal capacity S_{therm} is therefore the decisive factor here; naturally cooled cables laid in the ground therefore have only a low overload capacity. The "Natural Load" of high voltage cables is considerably greater than their thermal load capacity; the ratio is approximately 2:1 to 6:1, i.e a reversal of the ratio compared to overhead lines. Typical values for 110 kV and 380 kV are shown in Table 1.1. In total, therefore, the specific load and overload capacity of high voltage cables is comparatively lower than that of overhead lines. Cables are always operated below their natural load.

Table 1.1
"Natural load" P_{nat} and thermal load limits S_{therm} of overhead
lines and cables [3]

U_N	110 kV		380 kV	
System	O'head.	Cable	O'head.	Cable
P_{nat} in MW	35	255	600	2.900
S_{therm} in MVA	130	125	1.800	900
S_{therm} / P_{nat}	~ 4:1	1:2	3:1	1:3

- *Reliability*
  The reliability of the individual items of equipment play a decisive role in the security of the electrical supply. Because of their different types of construction, there are important differences between overhead lines and cables with regard to their operational reliability. A measure of reliability is the availability of the transmission path. The frequency and duration of failures are the decisive factors here.

  Because of the exposed situation of the conductors, disturbances such as arcing faults caused by lightning strikes occur more frequently in overhead lines than in cables. Nevertheless, these disturbances cause no permanent damage that would lead to the failure of the overhead line. Disturbances to cables, e.g. due to building work, are less common but nearly always cause permanent damage. Disconnection and subsequent repair are unavoidable because the cable insulation is destroyed at the fault location and cannot regenerate itself like the air in the case of an overhead line.

  Repairs of typical damage to an overhead line usually only take a few hours. Repairing the damage to a 110 kV cable, on the other hand, takes considerably longer. The repair of cable damage in the 110 kV network takes, on average, 60 to 120 hours.

  To summarize, although the frequency of failures in cables is lower than that in overhead lines, the duration of the failure is considerably longer. The availability of overhead lines is therefore correspondingly higher than that of cables. It is also operationally important that repairs to cables are more costly then repairs to overhead lines.

- *Installation costs*
  The lengths in which cables can be supplied are limited for reasons of manufacture and transport (weight, permissible road and rail clearances). Lengths of 400 to 800 m are possible, depending on the type of cable (single core or three core cable). As a result of this, relatively high costs occur for the fitting of joints and, if necessary, associated building work for joints. The costs of laying the cables (direct in ground, pulling into previously laid tubes, laying in accessible tunnels) are also considerably higher than the construction of an overhead line. The civil engineering costs depend mainly on the route (open country, city area) and amount to more than 50% of the total costs of a cable system.

  The relationships between the construction costs of overhead lines and cable systems become ever more in favor of overhead lines as the voltage increases. The cost relationships between high and extra high voltage cable systems and overhead lines for 3 voltage ranges between 110 kV and 765 kV were determined in an international study carried out by Cigré in 1994/95, in which 19 countries participated [4].

When this study was presented at the 1996 High Voltage Conference in Paris, the working group drew special attention to the problems of this kind of cost comparison and gave the following relationships between overhead lines and cable systems as a guide for the three most important voltage levels:

110 kV   1:7
220 kV   1:13
380 kV   1:20.

Those using these guide figures should bear in mind that specific transmission projects with different limiting factors always require detailed research and cost examinations taking the whole system into account. The extent to which the relationship of construction costs can deviate upwards or downwards in individual cases – taking everything into account – is illustrated by an example of a 380 kV cable system that was commissioned in Berlin in 1994. The cost comparison carried out in the planning phase resulted in a ratio of 1:5 for the construction of overhead line as compared to cable. In all cases, however, cable systems require higher expenditure for investment and operating costs than overhead lines.

Overhead lines or cables should only be regarded as opposites or competitors at a superficial level. They are really two different possible solutions for the same technical problem. Both systems have advantages and disadvantages and both are the right solution in the right situation.

The different properties of cables and overhead lines in terms of insulating material, dimensions, charging current demand, transmission capacity, overload capacity and length limitation must be considered in individual cases. A general judgement of the two types of transmission system cannot therefore be made from individual aspects without consideration of network interconnections and limiting conditions.

There are many cases in which only overhead lines make sense on technical grounds, for example in the case of long distances or very high power. Equally, in other cases, the use of cable is forced by external circumstances, for example in conurbations and large cities or in cases of difficult construction conditions.

In all other cases, however, a choice can be made between overhead line or cable. Especially during the planning of new transmission systems, the basic question regularly arises as to whether cable or overhead line is the most suitable for the particular application. If cost aspects are the most important consideration, the decision is clearly in favour of overhead line, the gap opening further not only with increase in rated voltage but also with the power to be transmitted.

As already discussed in Section 1.1, questions of countryside usage, environmental protection and environmental compatibility are becoming of increasing importance alongside technical and economic considerations. In the future, environmental compatibility checks will have to be carried out for large transmission system projects, not least due to the demands of that part of the public that is affected. It is also true that the transmission and distribution of electrical energy can only be assured in the long term if it it is compatible with the environment. The question "overhead line or cable" thus leads inevitably to a conflict of objectives between economic efficiency and environmental compatibility.

1.3.2 Proportion of Cables in the Public Power Supply Network

At the end of 1995, the total length of all the cables and overhead lines in the public power supply network in Germany amounted to 1.5 million km. Of this, about 1.2 million km was in West Germany, where the proportion of cable is 71%. Transmission lines in East Germany amounted to a total length of 290 000 km and, here, the proportion of cable is 48% [1].

The analysis for 1995 by voltage level is shown in Figure 1.2. From this, it will be seen that 66% of the length of public power transmission lines in Germany was cable. The proportion was 77% at the low voltage level and 62% at medium voltage. There were about 4750 km of cable in the public supply network at high and extra high voltage levels (about 100 km of this for $U_N \geq 220$ kV). This represents a share of about 4%.

In the 110 kV networks, about 6% of the total circuit length is cable. Most of the 110 kV cable is laid in the large city areas and only about 1/6 is in country areas. Cable is therefore principally used at low and medium voltages. Overhead lines dominate in the high and extra high voltage area for the technical and economic reasons explained previously.

Figure 1.3 shows the development of the West German power supply network over the last 35 years. The total length of transmission line increased by more than 100% from 555 000 km in 1960 to 1245 million km in 1995. The total length of cable systems increased by nearly 750 000 km. The amount of overhead line decreased by more than 50 000 km between 1960 and 1995; the percentage share fell from 75% to 29% during this period. The expansion and renewal of the power supply network in Germany thus took place predominantly using cable. This also applies to the 110 kV network in urban areas.

The study carried out by Cigré Study Committees 21 and 22 [4] shows that, in the high and extra high voltage sector, the international statistics indicate comparable ratios to those in Germany. Of a total of about 860 000 km of $U_N \geq 110$ kV transmission systems

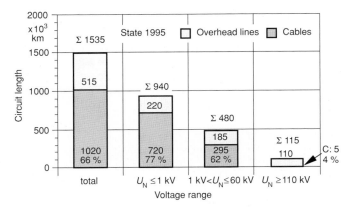

Figure 1.2
Lengths of transmission circuits in the German public supply network in 1995 (old and new Federal States) [1]

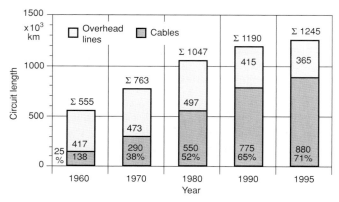

Figure 1.3
Development of the public supply network in Germany (only old Federal States) [1]

in the leading industrial nations (Western Europe, USA, Japan, Canada, Australia), only about 16 000 km, i.e. less than 2 %, were cable systems. Table 1.2 summarizes the percentage share of cable in existing, new and already approved and future planned transmission systems for three voltage levels. It can be clearly seen that the share of cable systems in the classic high voltage area will increase significantly in the future worldwide and that cable above 380 kV will only achieve a small share of the market.

In 50 projects evaluated in the study, the most important reason given for the choice of cable in the high voltage area was:

• Public debate   31 times

• Environment reasons   29 times

• Technical reasons   17 times

It can be seen from this that public debate and environmental aspects will also play a dominant role in other countries in the future in the choice of power transmission system.

Table 1.2
Proportion of cables in existing and planned transmission routes
in the leading industrial countries (status: 1994 [4])

Voltage range	Installed routes	Approved routes	Planned routes
110…219 kV	2.7 %	10.0 %	18.0 %
220…362 kV	1.0 %	1.7 %	4.0 %
363…764 kV	0.3 %	0.4 %	0.8 %

2 Types of Cable, Areas of Application and Components

This section provides a brief overview of the application, types and components of common high voltage and extra high voltage cables, i.e. cable with impregnated paper or extruded solid insulation. Gas insulated lines and low temperature cables are not taken into account at this stage (see Sections 5.2 and 5.3). Firstly, however, a brief outline is given of the historical development of cable technology.

2.1 Historical Development of High and Extra High Voltage Cable Technology

The basic invention of cable technology, the insulated conductor, can be dated at about 1830. It then took nearly fifty years until the first underground cable for the transmission of power was brought into use in Berlin in 1880. This was the worldwide beginning of *power cable technology*.

At first, the insulation materials available were gutta-percha, a type of plastic rubber, impregnated jute and, somewhat later, impregnated paper. Only the paper dielectric was finally able to fulfil the requirements of resistance to heat and of dielectric strength.

The development of high voltage power cable with paper dielectric is marked by five key technologies:

1. The introduction of the multi-layer dielectric using lapped paper tapes (Ferranti, 1880)

2. The vacuum impregnation of dried paper insulation using heated resinous mass (thus the description "mass-impregnated cable")

3. The moisture-proof encapsulation of the insulation by a seamless, continuously applied lead sheath (Borel, 1879)

4. The limitation of the electrical field of the cable cores using metallized paper to create the radial field cable (Höchstädter, 1913)

5. The impregnation of the paper dielectric with low-viscosity insulating oil under permanent pressure (the "oil-filled cable" principle from Emanueli, 1917).

The brilliant invention by Ferranti led not only to the introduction of an insulation material that showed the way to the future but, at the same time, used an insulation principle whose importance was recognized much later and is suitable for use up to the highest voltages: the multi-layer dielectric that distributes and reduces the risk of failure by using layering. This allowed Ferranti to produce the first single phase alternating current cable for 10 kV.

The second decisive milestone on the way to high voltages was set by Höchstädter. To improve the distribution of the electrical field, he introduced screening of the cores using metallized paper (*Höchstädter foil*), allowing the ionization threshold voltage of the cable to be increased considerably. This improved design (so-called *radial field cable*) allowed mass-impregnated cable up to a rated voltage of 60 kV to be produced.

At higher voltages, a further principal disadvantage of this design of cable becomes evident; the formation of voids within the insulation caused by change of thermal loading during operation. Partial discharges occur in these voids at correspondingly high field strengths, leading to the destruction of the dielectric.

The 100 kV barrier was first passed with the introduction of the fifth key technology, the oil-filled cable invented by Emanueli. The first *partial discharge free* and *thermally stable* cable for a rated voltage > 100 kV was commissioned in 1924 in a research facility in Italy. This was an oil-filled paper-insulated cable, as still used in a similar form today in high and extra high voltage ranges.

Thermal stability is understood as the ability of the insulation to adapt itself to the load changes that are unavoidable in operation and the changing temperatures resulting from this without the formation of voids, so that no ionisation processes occur even at the high operational field strengths that are usual in the high voltage range. Emanueli was able to do this by impregnating the layered paper insulation with a low-viscosity impregnation medium (insulating oil) which – under continuous positive pressure – flowed into the expanding dielectric from reservoirs on heating and was forced back into these on cooling.

This technology was then refined step by step to cope with ever increasing rated voltages and was available for 220 kV since the 1930s, for 400 kV since the beginning of the 1950s and even for 500kV since 1974. To demonstrate the feasibility of cables for even higher voltages, experimental oil-filled cables were tested successfully at 750 kV (1965 in France) and 1100 kV (1980 in Italy by Pirelli).

Parallel developments also led to thermally stable paper insulation using high-viscosity or even non-draining impregnation media. In these cases, ionisation was prevented either by compression from an external source (external gas pressure cable) or using gas or oil that penetrated the voids under high pressure (internal gas pressure or high-pressure oil-filled cable).

The desire of customers for the development of cables that were maintenance-free and easier to install finally led back to a single layer dielectric without impregnation medium, similar in principle to gutta-percha from the first days of cable technology. With the use of suitable synthetic plastics, it was possible to return to the simple, continuous process of extruding the insulation over the conductor and thus replace the paper wrapping technology. This solid extruded polymer insulation represented a completely different solution for the thermally stable high voltage cable. When correctly processed, this single-layer single-medium dielectric is intrinsically free of voids and partial discharges. Impregnation with a fluid component is therefore unnecessary.

The first precursor of the present day polymer-insulated cable was manufactured from PVC in the mid-1940s. However, this insulation material proved unsuitable for high voltage cables due to its high dielectric losses. The decisive milestone in the new high voltage cable technology was only reached with the use of thermoplastic polyethylene (PE) with its excellent electrical properties.

The largely non-polar PE was first used as a cable dielectric in the USA in 1944 at 3 kV and in 1947 in Switzerland at 20 kV. In 1966, the first high voltage PE cable was commissioned at 138 kV and 3 years later came the first 220 kV PE cable. Finally, in 1986, followed the first use of 400 kV cable with extruded thermoplastic insulation in France.

In the meantime, cross-linked polyethylene (XLPE), which is thermoelastic and is thus suitable for high operating temperatures (90 °C instead of 70 ... 80 °C), has fully displaced the thermoplastic variant as cable insulation worldwide. It was first used in high voltage cables at the beginning of the 1960s, and XLPE insulated cable up to 500 kV has been available since 1988. The first 400 kV XLPE cables in Germany were installed in 1997.

The development of high voltage cable with extruded dielectric is marked by the following key technologies:

1. The extrusion of single-layer insulation using thermoplastic polyethylene

2. The extrusion of the insulation and semiconductive layers in one operation using a triple injection head

3. The moisture-proof metallic sheathing of the insulation

4. The use of cross-linked polyethylene as insulation material

5. The cross-linking of PE in a nitrogen atmosphere (*dry curing*)

6. The use of insulation compounds of highest purity and the manufacture of the dielectric under clean-room conditions

Table 2.1 Historical development

1.	Beginning of cable technology
1847	First communication cable with gutta-percha insulation
1880	First power cable with gutta-percha insulation (for DC))
1882	First power cable with impregnated textile insulation and lead sheath
2.	Power cables with paper dielectric
1890	10 kV mass-impregnated cable (for AC)
1913	33 kV mass-impregnated cable (radial field cable)
1924	132 kV low-pressure oil-filled cable (thermally stable cable)
1931	External gas pressure cable
1931	High-pressure oil-filled cable(Oilostatic cable)
1936	220 kV low-pressure oil-filled cable
1937	Internal gas pressure cable
1952	400 kV low-pressure oil-filled cable
1974	500 kV low-pressure oil-filled cable
1980	1100 kV low-pressure oil-filled cable (experimental cable)
3.	Power cables with extruded dielectric
1947	20 kV cable with PE insulation
1960	20 kV cable with XLPE insulation
1966	138 kV cable with PE insulation
1969	225 kV cable with PE insulation
1979	275 kV cable with XLPE insulation
1986	400 kV cable with PE insulation
1988	500 kV cable with XLPE insulation

The development of high and extra high voltage cable, especially paper-insulated cable, was then oriented towards being able to offer cables for the particular rated voltages of the individual network levels. From about 1960, the increase of current-carrying capacity of the existing cable types to cater for high-power transmission became an important direction of development. The main task here was the dissipation of the considerably higher current related losses.

Considerable increases in transmission capacity were achieved in conventional cables using various methods of forced cooling. Forced-cooling technology was developed more and more in several countries – especially in England, Germany and Japan – and in various ways. The highest power that could be transmitted using these methods was about 2500 MVA.

For underground high-power links from about 2000 MVA, new types of cable such as gas-insulated or superconductor cables have been developed or are under development.

Table 2.1 summarizes the most important steps of development discussed here in the form of a list.

2.2 Types of Cable and their Areas of Application

2.2.1 Three-Phase AC and Direct Current Cables

Electrical power can be transmitted as alternating or direct current. In meshed networks with different voltage levels in a relatively narrow space, as is the case in the power supply systems in most countries of the world, three-phase alternating current at a frequency of 50 or 60 Hz has become predominant because it is easily adapted to various voltage levels by using power transformers. As a result, underground transmission routes are also operated with three-phase AC. The associated cable systems consist of three cables whose phase to ground voltages have a mutual phase displacement of 120° in stable operation.

High voltage cables closely resemble very long cylindrical capacitors whose capacitance per unit length C' is between about 120 and 600 nF/km depending on dielectric, conductor cross-section and thickness of insulation. Therefore, under AC load, they draw a capacitive charging current I_c, amounting to

$$I_c = U_0 \, \omega \, C' \, l \propto U_0 \, \omega \, \varepsilon_r \, l \tag{2.1}$$

per phase. In the above equation, U_0 is the phase voltage, ω the angular frequency, ε_r the dielectric constant and l the cable length. The maximum charging current at the start of the cable resulting from the equally distributed capacitance per unit length is superimposed on the actual consumer current that is to be transmitted along the route. As a result of this, there is a theoretical critical length l_c for a three-phase cable at which the maximum thermally permissible operating current is already exceeded at the feed point at no load because of the capacitive load of the cable. Thus, l_c reduces in accordance with Equation 2.1 with increasing voltage and permittivity of the insulation for a given type of cable (material, cross-section, design, see Section 2.3.1)[1].

---

[1] The limiting length defined here is actually even shorter than the value calculated from I_c alone, as a further voltage-proportional component to cover the dielectric losses is added vectorially to the capacitive charging current and effective current.

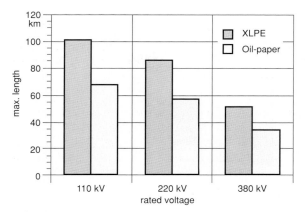

Figure 2.1
Max. transmission lengths for naturally cooled three-phase cable systems with 1000 mm² Cu conductor utilizing at least 80 % of the full-load current for the consumer

Economically worthwhile operation of a cable is only possible if the system installed can be largely utilized to supply power to the consumer. A guide value is that at least 80 % of the current fed into the cable should be available to the consumer at the end of the cable. Figure 2.1 gives an overview of the resulting transmission length for naturally-cooled single-core cables with 1000 mm² copper conductor and oil-filled paper or XLPE insulation. The approximately 50 % greater lengths for XLPE cable result from the considerably lower permittivity of 2.3 compared to 3.5 for oil-filled paper.

While the lengths shown in Figure 2.1 are usually sufficient for the meshed high voltage cable networks in heavily populated areas, the limits are quickly reached in, for example, submarine cable links. Since overhead lines are out of the question here, the only remaining option is to use DC cables. In the case of DC ($\omega = 0$) there is no capacitive charging power; also, the additional losses caused by e.g. skin effect and induced sheath current are not present. In addition, unlike the situation with 50 Hz, the voltage-dependent losses in the dielectric can be practically neglected (see Section 3.4). As a result, DC cables offer considerable advantages for long transmission routes, and practically all the long submarine cable routes are operated with DC for this reason. Nevertheless, converter stations are required at both ends of the transmission route for the conversion from AC to DC and vice versa.

2.2.2 Design Features

High and extra high voltage cables have common design features – independent of the type of dielectric, the rated voltage and operating frequency – that will be examined briefly at this stage as the basis for more detailed treatment in subsequent sections (see Figure 2.2). The components that essentially determine the electrical and thermal behavior of the cable are the conductor, the insulation with inner and outer field limitation and the metallic sheath.

Cables of the voltage ranges being considered here are designed as so-called *radial field cables*. This means that the conductor with screen and the metal sheath form a long con-

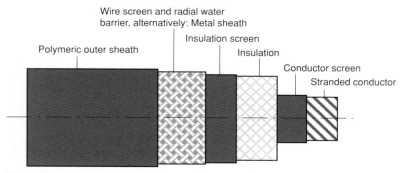

Figure 2.2 Principal design elements of common high and extra high voltage cables

centric cylinder and that the dielectric forming the *insulation layer* is electrically stressed exclusively cylinder-symmetrically when looked at macroscopically, in accordance with Equation 2.2:

$$E(x) = U_0 / [x \cdot \ln(R / r)] \qquad (2.2)$$

where U_0 is the operating voltage, R and r are the external and internal radii of the insulation layer and x is the continuous coordinate.

This statement also applies to external gas pressure cables with slightly oval conductor contour (see Sections 2.3.1 and 4.1.5), as the ovality is insufficient for a significant influence on the field distribution. To ensure that the radial field remains undisturbed, even in the areas close to the stranded conductor and screen and to the corrugated metal sheath, so-called stress-relieving layers are provided between these elements and the insulation (see Section 2.3.3 etc. for details). The grounded metallic sheath, which is always needed, provides effective electrical screening of the cable. The cable environment is thus free of *electrical* fields. The cable is finally given an overall sheath of suitable thermoplastic material to protect the metallic sheath from corrosion damage.

Apart from gas-pressure cables with three individually screened cores in a common steel pipe, the three individual phases of a three-phase high voltage cable system are generally laid as single core cables in various configurations direct in the ground or in tunnels in free air. Only DC cable systems have two, or in some cases just one live conductor; in the latter case – for submarine cables – the surrounding sea water is used as the return conductor.

2.2.3 Cable Dielectrics

Two different groups of dielectrics are available as insulation for the high and extra high voltage cables being considered here: impregnated paper and extruded synthetic materials.

Impregnated paper

As a natural product from selected softwoods, paper has been used for the insulation of power cables for more than 100 years. Depending on the rated voltage, the dielectric consists of a few tens (60 kV) up to several hundreds (≥ 400 kV) of narrow strips lapped

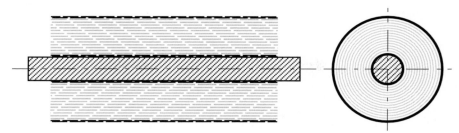

Figure 2.3 Principle of construction of lapped paper insulation

spirally over each other, the butt gaps of these being axially displaced from layer to layer (see Figure 2.3). There are therefore inevitably air spaces in the axial gaps and between the individual machine spun paper layers (essential in the case of the gaps to allow the cable to be bent – see Section 4.1). Furthermore, the paper itself is not a homogeneous material; in fact only about 50 to 60% of the volume consist of cellulose molecules, the remaining pores contain largely air in the as-supplied state.

To effectively prevent pre-discharges in the electrical field within these spaces, the air must be removed and replaced by an electrically stable medium. The only materials suitable for this are those that can flow, at least during processing, in order to be able to penetrate the gaps and fill these completely. According to the type of cable, the materials used for this purpose are either permanently low-viscosity insulating oils (low-pressure oil-filled cable, LPOF) or so-called impregnation mass, which is only of low viscosity at processing temperature, but is highly viscous or even solid under operating conditions (high-pressure oil-filled and gas-pressure cable).

The most important common feature of all paper-insulated high and extra high voltage cables is the multi-layer construction of the dielectric. Differences result mainly from the type of impregnation medium used to prevent pre-discharges. This type of insulation thus has a multi-layer multi-material dielectric.

An exception among impregnated paper dielectrics is a development for the highest transmission voltages: the so-called plastic foil paper laminate. Here, the paper is coated with a foil – polypropylene (PP) among others – and can be processed to form multi-layer cable insulation (*P*olypropylene *P*aper *L*aminate PPL) in the same way as the usual cellulose paper. Only low-viscosity synthetic insulating liquids are used as impregnating media, as mineral oil can cause swelling of the polypropylene part of the laminate. The main advantage of the more costly PPL insulation compared with the usual paper dielectrics is its considerably lower dielectric loss and greater electrical strength, thus allowing higher electrical operating field strengths (see Sections 3.4.2 and 4.1 for further details).

Extruded synthetic dielectrics (PE, XLPE, EPR)

The costly processing of paper and impregnating medium to form a voltage-resistant impregnated multi-layer dielectric, consisting of several individual steps, some of these time-intensive, favored the development of a solid dielectric that could be extruded over the conductor in a continuous, fully automatic operation that, in addition, required no impregnating medium. The prerequisite for the use of such a single-layer dielectric in high voltage cable technology was the availability of low-loss and permanently voltage-resistant materials that could be extruded without voids. These conditions are fulfilled by the naturally non-polar thermoplastic polyethylene (PE) and its chemically cross-linkable variant XLPE which, because of its thermoelastic properties, permits higher operating temperatures and has large thermal reserves in the case of short circuits. The insulation of an extruded cable is therefore based on a single-layer, single-material dielectric.

Both polymer materials and their processing technologies have been continuously developed since the production of the first PE-insulated high-voltage cable in the 1960s and are now available as dielectrics for the entire voltage range up to 500 kV. In addition to improvements in the purity of the material, a decisive step forward was the introduction of triple extrusion enabling the semiconductive layers at the boundary surfaces of conductor and screen to be welded solidly to the insulation during production (see Section 4.2). In the meantime, XLPE has replaced PE as the insulation medium for high and extra high voltage cables in almost all countries.

Apart from XLPE, synthetic rubber (*E*thylene *P*ropylene *R*ubber, EPR) has also been adopted to some extent in certain markets (e.g. Italy, South America) as an extrudable single-layer dielectric for high-voltage cables. Nevertheless, EPR is limited to the voltage range up to about. 150 kV because of its comparatively high dielectric losses. It has, however, good thermal properties, high elasticity and, compared to XLPE, less susceptibility to partial discharges.

In Germany, neither EPR nor thermoplastic polyethylene come into question. Newly installed extruded insulated cables for ≥ 60 kV here have – as in most other countries – exclusively XLPE insulation.

2.3 Components

The principal components of the common types of high and extra high voltage cable have already been examined briefly under the heading "Design Features" (Section 2.2.2, Figure 2.2). The most important properties and the function of these components will now be described insofar as they apply in general to the types of cable being discussed here.

2.3.1 Conductor

The task of the conductor is to transmit the current with the lowest possible losses. The decisive properties for this function result in the first place from the conductor *material* and *design*. The conductor also plays a decisive part in the mechanical tensile strength and bending ability of the cable.

Conductor materials

In the German Federal Republic, high and extra high voltage cables are equipped almost exclusively with copper conductors while, in other countries, e.g. France, aluminum (Al) is also widely used as a conductor material. The advantages of copper (Cu) are its approximately 60% lower specific resistance and the resulting smaller conductor cross-section for a given current carrying capacity, the correspondingly smaller diameter affecting all the subsequent components of the cable. In addition, copper has about three

Table 2.2 Properties of Al and Cu conductor materials

Property	Copper	Aluminum
Density in g/cm^3	8.89	2.70
Spec. resistance in Ωmm^2/m	$17.24 \cdot 10^{-3}$	$28.26 \cdot 10^{-3}$
Tensile strength in N/mm^2	200...300	70...90

times the tensile strength of aluminum, which shows advantages in the processing of the conductor. The principal advantage of aluminum lies in its lower density, some three times less than that of copper, leading to practically half the weight at the same cable capacity. Table 2.2 compares the most important properties of the two conductor materials. In addition, it should be mentioned that the electrical properties, especially, are strongly influenced by the purity and the thermal-mechanical processing (compression, annealing) of the material.

High rated currents and high transmission voltages and the insulation thicknesses required to withstand these clearly favor copper conductors because of the smaller overall cable diameter thus permitted, allowing more economic transport lengths per drum. On the other hand, the lower current and voltage ranges can be covered just as effectively with Al conductors, as confirmed by examples in other countries. There are no serious technical problems in the use of Al as conductor material. The main criterion nowadays for the choice of conductor material is an economic one.

The cross-sections of cable conductors are standardized. The nominal cross-sections for the high and extra high voltage range are between 240 mm^2 to 2500 (3000) mm^2 for copper conductors. The upper limit is primarily determined by the bending ability and weight of the cable.

Conductor design

Two parameters largely determine the design of the conductor: the value of the rated current and thus the cross-section required for its transmission and the type of dielectric. The principles of the individual designs are shown in Figure 2.4.

The dielectric determines the conductor configuration in low-pressure oil-filled cables, which basically have a hollow conductor that allows the low-viscosity impregnating medium to flow to and fro under the influence of thermal load change.

Another design that is determined by the insulation medium is the oval conductor used in external gas-pressure cables. In this case, the oval conductor contour facilitates void-free adaptation of the high-viscosity impregnated paper insulation under the influence of the gas pressure acting on the core from the outside under various load conditions.

All other design variants – including variants of the hollow and oval conductors already mentioned – are adapted to the necessary conductor cross-section taking manufacturing and/or economic aspects into consideration. The selection criteria here are the bending ability of the conductor and the cable, the diameter and stability of the strand bundle in the case of stranded construction and, above all, the effective conductor resistance with

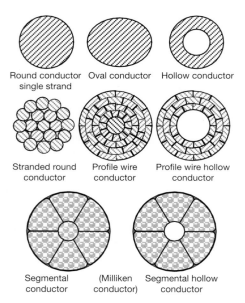

Figure 2.4
Conductor designs for
high-voltage cables

mains frequency alternating current, which, unlike the DC value, rises disproportionately with increasing conductor diameter under the influence of skin and proximity effects. Table 2.3 provides an overview of the most important features in this respect of the conductors shown in Figure 2.4 and indicates their preferred areas of application.

Table 2.3 Properties of various types of conductor

Conductor/brief descr.	Advantages	Disadvantages	Application
Single-strand conductor/RE (solid conductor)	• 100 % filling factor • Homogeneous surface • Low cost	• Not flexible • Considerable current displacement	Up to max. 800 mm² in Al version
Stranded round cond. compacted/ RM(V)	• Flexible • Simple construction	• Current displacement • Risk of collapse points with extruded dielectric	Cross-sections up to approx. 630 mm²
Profile wire cond./ / RMF, also as hollow conductor RMF/H ("Conci conductor")	• Good filling factor • Stable bundle • Homogeneous surface	• Less flexible • Increased current-displacement	For extruded cables from 800…1000 (1200) mm², up to 2500 mm² as hollow conductor for oil-filled cables
Segmental conductor/ RMS ("Milliken conductor") with 4, 5 or 6 segments, also as hollow conductor/RMS/H	• Very flexible • Reduced current-displacement, especially with individually insulated wires	• Costly • Poor filling factor • Risk of collapse points with extruded dielectric	Large cross-sections from appr. 1200 mm², also as hollow conductor for oil-filled cables
Stranded oval conductor/ OMV (not illustrated)	See stranded conductor	See stranded conductor	External gas pressure cables

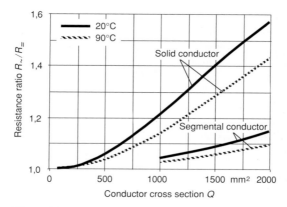

Figure 2.5
Increase of conductor resistance due to skin effect at rated current and 50 Hz

It can be seen from this that stranded conductors are used almost exclusively, as solid copper conductors of the cross-sections normally used for high and extra high voltage cables are too rigid. An exception is the single strand circular soft aluminum conductor that is used for cross-sections up to 800 mm². The bending ability of a stranded conductor depends mainly on the number of individual wires and the pitch of the stranding. To achieve a good filling factor and a stable stranded bundle, stranded conductors are compacted to a greater or lesser degree during processing using suitable tools. However, the compacting leads to more intensive axial contact between the individual wires, so that a heavily compacted stranded conductor is not noticeably better than a solid conductor in terms of *current displacement* due to *skin effect* and *proximity effect*.

As an example of the effect of current displacement due to skin effect alone, Figure 2.5 shows the relationship between AC and DC resistance at 50 Hz against the cross-section of a solid copper conductor at rated current and at 20 °C and 90 °C[1]. Under these conditions, the AC resistance $R_\sim$ of the conductor at 90 °C increases by nearly 45 % compared to the ohmic resistance $R_=$ for the largest cross-sections considered here. Lower initial resistances actually lead to even poorer utilization of the cross-section at 50 Hz, as shown by the example for 20 °C with its some 30 % lower ohmic resistance.

A much more favorable relationship of $R_\sim / R_=$ results from the use of a segmental conductor instead of the solid conductor: the rise is only 10 to 15 % depending on conductor temperature. For this reason, large cross-sections are now preferably configured as segmental conductors (so-called *Milliken conductor*). With this design of conductor, the segments are insulated from each other to constrain the current flow. The boundary between the two designs depends on, among other things, the conductor material and its specific ohmic resistance and on whether or not an oil channel is required. In the case of copper conductors without oil channel, as used for polymer insulated cable, a Milliken design is recommended for cross-sections above about 1200 mm²; for less conductive aluminum (see Table 2.2) and hollow copper conductors, in which the effect of current

---

[1] The case "Rated current at 20 °C" is a theoretical assumption and demonstrates the effect on the current displacement of the approx. 30 % higher material conductivity. In practice, the rated current naturally leads to the conductor heating up to rated operating temperature.

displacement is less because of the greater diameter, the limit usually lies one nominal cross-section higher. The advantage of segmented design is only fully realised if – as is usually the case – the segments are not only insulated from each other as a whole, but the individual wires within the segments are also insulated from each other (the graph in Figure 2.5 is based on this case). Suitable processes for this are lacquering or oxidation of the wire surface. Without *individual wire insulation*, the effect depends largely on the actual boundary resistance between the individual twisted wires. A high degree of compacting, normally desirable to achieve a stable construction and good filling factor, is counter-productive in this respect. In worst cases, the advantages of the Milliken design are virtually nullified. In practice, the AC resistance of segmented conductors lies approximately in the middle between the graphs of the solid conductor and the ideal Milliken conductor in accordance with Figure 2.5 [5, 6], depending on the construction. Further information about current displacement is given in Section 3.4.1.

2.3.2 Insulation Layer

The insulation layer of the cable has the form of a hollow cylinder and consists of a dielectric material or material combination. As already discussed in 2.2.3, the only practical dielectrics for the common types of high and extra high voltage cables are the two main groups "impregnated paper" as a multi-layer arrangement or "extruded polymer" as a single layer insulation. From the high voltage technology point of view, the insulation layer is the most important component of the cable and the most difficult one to calculate in terms of its long-term behavior. Questions of the electrical rating and realistic service life forecasts for the dielectric with expected operating periods of 30 to 40 years are therefore among the most important tasks in the development and design of high and extra high voltage cables.

There has now been nearly seventy years experience in the use of the classic multi-layer paper and insulating oil dielectric, during which period impregnated paper has proved exceptionally reliable and durable. With correct electrical and thermal design (Sections 3.2 and 3.1.2 respectively), there is no need to worry about *ageing processes* in either the paper or the impregnating medium during the expected operating period.

The considerably shorter history of high voltage cable with extruded solid dielectric was marked by a series of setbacks at the beginning that, at first, stood in the way of the wide application of this technology in the high voltage area. The necessary reliability and long-term stability of solid insulation, comparable with that of paper dielectric, was only achieved by co-operation between the cable manufacturers and the polymer suppliers to provide better material purity, the optimization of manufacturing technology to achieve homogeneous boundary layers between the insulation layer and the field-limiting semiconductive layers and the effective exclusion of moisture during the manufacture, laying and operation of the cable [7].

2.3.3 Field Limiting Layers

To ensure a defined cylindrical field and to withstand the field strengths that occur, all cables for $U_N > 6$ kV, independent of their type of dielectric, require field limiting or field smoothing layers, widely known as semiconductive layers, in the interface between conductor and insulation and between insulation and metal sheath, and these layers are specified in the standards:

- Inner semiconductive layer ("conductor screen").

- External semiconductive layer ("core screen").

These have two principal purposes:

- Equalization and reduction of the electrical stress in the cable dielectric by preventing local field enhancement in non-homogeneous areas such as the individual wires of the conductor or screen. The semiconductive layers eliminate the effect of the individual wires on the field distribution (see Figure 2.6).

- Prevention of the formation of gaps or voids between the voltage-carrying components of the cable (conductor, screen and metal sheath) and the insulation layer due to mechanical stress, e.g. bending of the cable or differential expansion of the various materials under varying thermal stress. A solid and permanent bond between the semi-conductive layers and the insulation effectively prevents the occurrence of partial discharges; an essential feature in the case of polymer-insulated cables, which have no impregnating medium.

The first-mentioned function can, in principle, be fulfilled by any sufficiently conductive material, provided it is compatible with the insulation and is able to permanently form smooth and homogeneous boundary surfaces with the insulation.

On the other hand, effective permanent prevention of the formation of gaps between the insulation layer and the field smoothing layers under heavy varying thermal stress is only ensured by using layer materials whose coefficient of thermal expansion is as close as possible to that of the dielectric being used. As a result, largely the same material is chosen for the field smoothing layers as for the insulation: paper tape (*carbon paper*) in the case of paper-insulated cables and extrudable polymers (*conductive compounds*) in the case of cables with solid dielectric. The necessary conductivity is created by the incorporation of special carbon black in the carrier material. Paper tape coated with aluminum on one side is also used for paper-insulated cables.

In the case of impregnated paper dielectrics, the problem of the formation of thermally caused voids between the semiconductive layer and the insulation layer is considerably less than that with extruded polymer, since the microvoids that exist in any case

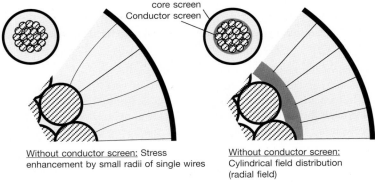

Without conductor screen: Stress enhancement by small radii of single wires

Without conductor screen: Cylindrical field distribution (radial field)

Figure 2.6
Principle of field equalization over a stranded conductor by using a conductive layer

31

between the individual layers of the multi-layer dielectric are immediately filled by the flow of impregnating medium or pressurizing gas when expansion takes place. Any temporary partial discharges that may occur are extinguished again before permanent damage can occur (see Section 4.1 onwards). This kind of self-healing effect cannot occur in polymer cables that have no impregnating medium and, for this reason, the permanent bond between the semiconductive layers and the insulation has a much greater importance. This applies especially to PE and XLPE, which are very sensitive to partial discharges. For this reason, high and extra high voltage cables with extruded dielectric nowadays only have semiconductive layers that are permanently bonded to the insulation. In addition, to prevent the inclusion of foreign matter, these are applied to the conductor by *triple extrusion*, i.e. all three layers are extruded in one process. Further details of the properties, composition and processing of conductive compounds can be found in Section 4.2.

Besides the requirement for a permanent void-free bond, the ease of removal for the fitting of accessories must also be considered in the case of the outer semiconductive layer.

2.3.4 Metallic Covering

The term "metallic covering" applies to the various types of outer conductor that surround the cable core concentrically, i.e. cable screens and metal sheaths, also thin aluminum foils that are often used in polymer cables as part of a so-called laminated sheath that performs the function of a metal sheath together with the wire screen. In addition to the task of electrostatic screening already mentioned, the metallic covering also has to fulfil the following functions:

- Return of the capacitive charging current under operating conditions in accordance with Equation 2.1 without exceeding the permissible voltage drop along the metallic covering;

- Conduction of the earth fault current in the case of a fault until the system is switched off without exceeding the maximum permissible short-circuit temperature for the particular type of cable ;

- Reduction of the electrical influence on the cable surroundings in the case of a fault, e.g. an earth fault;

- Provision of protection against accidental contact;

- Protection of paper-insulated cables against leakage of impregnating medium;

- Mechanical protection of insulation while allowing the cable to bend sufficiently.

The first three requirements, especially the behavior in case of fault, require a definite minimum cross-section of the metallic serving; this is also influenced by a number of additional parameters and can thus only be defined exactly for a specific application case ([8], Page 281 ff.). Influencing factors in this case are, in addition to construction and materials, the serving itself including cross-section and rated current, short-circuit capacity, earthing and protection relationships in the network, the laying pattern of the cable system and its environment, etc.

The metallic covering can use copper as screening wires and lead, aluminum and steel as the sheath, steel only being used in the form of prefabricated pressure pipes welded together on site as a common container for the three phases of gas pressure or high-pres-

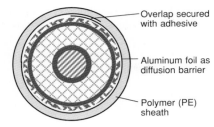

Overlap secured with adhesive

Aluminum foil as diffusion barrier

Polymer (PE) sheath

Figure 2.7
Laminated sheath for polymer-insulated cable with axially-glued Al foil as diffusion barrier

sure oil cable systems. The individual core screening also necessary to produce the radial field in these types of cable is carried out here using copper tapes or lead sheathing.

The necessary flexibility and bending ability of the metallic covering – with the exception of the pressure pipes just mentioned – is ensured when using copper wire screens and lead sheaths by the design (of the screen) and softness of the material (Pb). Sheathing layers made from thin aluminum foils laminated with polyethylene can also follow the smallest radius specified in the applicable standards without damage. The important factors here are, above all, good adhesion of the aluminum foil to the polymer sheath and firm glueing of the axial foil seam to prevent separation or leakage under bending stresses (see Figure 2.7).

On the other hand, solid aluminum sheaths – at least for cable diameters common in the high and extra high voltage range – must be produced with a defined corrugation, otherwise they would kink or split as soon as they are wound onto the transport drums. Two different methods are available for producing these *corrugated aluminum sheaths*; *spiral corrugation* or *ring corrugation*. The differences are illustrated in Figure 2.8.

The ring corrugation has the advantage that the cable can easily be made *longitudinally watertight* by the use of suitable expanding filler material over the core in the same way as lead or laminated sheath designs.; any moisture that penetrates due to damage is restricted to the immediate surrounding of the leak. On the other hand, spiral corrugation specifically allows liquid to flow in an axial direction, an advantage for low-viscosity impregnated paper insulation.

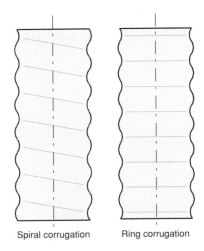

Spiral corrugation   Ring corrugation

Figure 2.8
Types of corrugated aluminum sheath

Table 2.4 Types of metal sheath

Type of sheath	Advantages	Disadvantages	Application
Cu-wire screen with Al laminated sheath	• Light in weight • Small cable-diameter • Longitudinally watertight (LWT)	• Limited protection against mechanical damage	• Polymer-insulated cables
Lead sheath	• No corrugation required • Small cable-diameter • Easily made LWT	• Heavy • Pressure protection needed with oil-filled cables	• Polymer-ins. cables • Low-press. oil-filled cables • Single cores of external gas pressure cables
Corrugated aluminum sheath	• Light weight • Good mech. protection • High short-circuit-current capacity	• Large cable diameter because of corrugation • LWT difficult with spiral corrugation	• Polymer-ins. cable • Low-press. oil-filled cables • Single-core gas pressure cables
Steel pipes	• Best mech. protection • Pipe laying independent of cable laying • Favorable reduction-factor	• Rigid • High installation-effort	• Gas pressure cables • High-pressure oil-filled cables

Whatever the type of corrugation, polymer-insulated cables must have a defined gap with sufficient padding between the core and the aluminum sheath to prevent damage to the insulation layer on heating caused by the high coefficient of expansion of most polymers and yet be able to fix the core firmly when cold. This measure is not needed for lead or laminated sheaths because of the natural expansion ability of the materials used. To permanently withstand the internal operating pressure of several Bar, oil-filled paper-insulated cables with lead sheath must have an additional pressure protection tape consisting of hard copper or stainless steel tape over an insulating foil applied over the lead sheath. In general, laminated sheaths are not used for impregnated-paper insulated cables.

Table 2.4 provides an overview of the various types of metallic sheathing with advantages and disadvantages.

2.3.5 Outer Coverings and Corrosion Protection

With the exception of the laminated sheath design, in which the Al foil used as a diffusion barrier forms a homogeneous whole with the polymer sheath, metal sheaths require additional protection against mechanical damage (for example while laying the cable) and, above all, against corrosion caused by water in conjunction with electrolytically active components in the soil. An external sheath of high-density polyethylene (HDPE) provides good mechanical protection and excellent resistance to abrasion together with low moisture penetration, and, as a result, HDPE has largely displaced the previously common PVC. Only in the case of lead sheaths, which have lower susceptibility to corrosion because of the material, is low-cost PVC still used as outer covering as in the past.

The moisture-proof property of the polymer is not sufficient on its own to permanently prevent any possible corrosion damage to the metal sheath. As a result, so-called *passive corrosion protection* must be provided before the actual external sheath is extruded over the cable. Optimum corrosion protection, good adhesion to the metal and chemical resistance is provided by high-viscosity bitumen-based compounds in conjunction with textile tapes as carrier material.

In the case of the steel pipes required for gas-pressure cables, it is recommended, if necessary, to supplement these passive protection measures with so-called *active corrosion protection* (cathodic protection). This involves the application of a defined DC voltage to the steel pipe of about $-1\,V$ with respect to ground, thus preventing metal-removing ions from flowing from the metal into the ground [9].

3 Physical Principles

3.1 Types of Stress in the Cable Dielectric

3.1.1 Electrical Stress and Field Distribution with Alternating Current

The electrical stress in the insulation results from the instantaneous phase-to-ground voltage and the geometry of the insulation layer. However, the macroscopic cylindrical symmetry of the radial field cable specified according to definition does not completely define the stress condition that actually occurs locally. To be more precise, the local field distribution in the insulation layer depends not only on whether the cable is being operated with AC or DC but, in the AC cable now being discussed in this section, also on the type of dielectric.

At first glance, homogeneous *solid dielectrics* such as PE, XLPE and EPR show no reasons to deviate from a cylindrical field. Nevertheless, basic investigations on suitable model arrangements and on cable samples have shown that, even here, noticeable field distortions that may reduce the breakdown strength of the insulation must be reckoned with. Two phenomena are basically the cause of this: microscopic *inhomogenities* in the dielectric and in its boundary surfaces with the semiconductive layers and the formation of *space charges*.

- *Inhomogenities*
  The influence of inhomogenities on the field distribution and, especially, the possibly resulting danger for the long-term behavior of the insulation are determined by the type of impurity, its form, size and location. A distinction is made between *inclusions* and microscopic voids (microvoids) in the dielectric and *protrusions* into the semiconductive layer boundary surfaces (see Figure 3.1). Within the group of inclusions, a distinction is made between conductive (metallic) particles and non-metallic particles, known as ambers (protrusions are, by definition, basically conductive impurities).

In terms of electrical stress, conductive impurities including semiconductive layer protrusions are of the greatest importance as, with unfavorable contour and position (pointed, radial direction) they can lead to theoretically infinite field magnification.

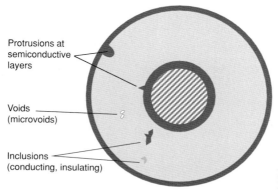

Protrusions at
semiconductive
layers

Voids
(microvoids)

Inclusions
(conducting, insulating)

Figure 3.1
Defects in polymer-insulated cables

Metallic inclusions present an additional risk factor in that, because they have a many times smaller coefficient of thermal expansion compared to the surrounding polymer dielectric, they may loosen and form microvoids in an already overstressed area. The prevention of even the smallest metallic inclusion therefore has the highest priority in the handling of material and the manufacture of solid-insulated high and extra high voltage cables [10]. A similar situation applies to protrusions, which are mainly avoided by the use of selected, especially homogeneous conductive compounds [11].

On the other hand, non-conductive particles consisting of oxidized or scorched insulation compounds are considerably less dangerous, as they only have a slight effect on the field distribution. Nevertheless, this kind of irregularity should be prevented as far as possible, as they have different thermo-mechanical properties from the surrounding material and therefore – in a similar way to metallic inclusions – can contribute in the long term to the formation of microvoids. To avoid problems, at least the size of the ambers is limited to definite maximum values that are nevertheless not yet specified for high and extra high voltage cables.

Microvoids with diameters of a few tens of μm are a typical phenomenon of steam cross-linked XLPE cable. Although they only have a slight effect on the field distribution in the surrounding dielectric, they are themselves, when filled with gas, stressed with a field strength higher by a factor of nearly ε_r than that in the region of the impurity. Therefore, taken together with the reduced dielectric strength of the gas filling, there is a risk of partial discharges in the microvoid. Following the introduction of dry cross-linking (see Section 4.2.3) and sensitive measurement of partial discharges during routine testing of each shipping length, "dangerous" microvoids no longer present problems in modern polymer-insulated cables.

A common feature of all the impurities discussed here is the fact that their importance is reduced as their dimensions become smaller. Microvoids lose their ability to ignite according to *Paschen's law* and field magnification caused by conductive inclusions or protrusions affects only microscopically small areas of the insulation [12] which, because of their minimal volume, exhibit an extremely high dielectric strength based on the so-called *volumetric effect*. As a consequence of this knowledge, the dimensions of the various impurities are limited, at least in manufacturers' own internal specifications, to maximum values that are regarded as uncritical for the type of dielectric and the macroscopic operational field strength. For conductive inclusions in extra high voltage XLPE cables, the upper limit is e.g. of the order of ≤ 50 μm. The contour also plays a decisive part in the case of protrusions, more exactly the ratio of height to width (h/b) of the almost semi-elliptical impurity as shown in Figure 3.2: the greater h/b, the stronger the field magnification; h/b values ≤ 1 count as relatively uncritical as, by calculation, the local stress is magnified by a maximum factor of 3. Table 3.1 provides an additional overview of the impurities and their effect on the electrical stress in polymer-insulated cables.

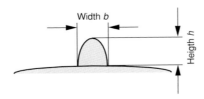

Figure 3.2 Idealized semiconductive layer protrusion

Table 3.1 Inhomogeneties and their effects in cables with solid dielectric

Type of impurity	Field magnification	Risks	Quality objective
Conductive inclusion, fine metallic particles	Any value, depends on contour	• Local electrical overstressing of the dielectric • Formation of microvoids on heating • Partial discharges	Max. size < 30 to 50 μm (Depends on U_N)
Non-conductive inclusion (amber)	None	• Formation of voids on heating • Partial discharges	Max. size < 50 to 80 μm
Microvoid	In microvoid $\varepsilon_r \times E_0$	• Partial discharges in microvoid	Max. size below detectable limit
Conductive layer-protrusion	Any value, depends on contour	• Local electrical overstressing of the dielectric	Max. size < 40 μm, $h/b \leq 1$

- *Space charges*

  Space charges, i.e. a collection of largely stationary charge carriers in the insulation, create an additional electrical source field that is superimposed on the geometrically determined field pattern and thus changes the resultant stressing of the dielectric. The main factors for the type and extent of the stress changes are the polarity, local distribution and concentration of the space charge.

  As extremely weakly conductive materials such as polyethylene and XLPE with insulation resistances above 10^{16} Ωcm usually tend towards electron enrichment, an excess negative charge gradually builds up in the dielectric [13, 14]. The charge carriers from the respectively negative electrodes (instantaneous cathodes) are injected into the dielectric under the influence of the locally prevailing electrical field and the current temperature and are stored at first in the immediate vicinity of the boundary surface. After the change of polarity, this same process takes place at the opposite pole; thus, what are known as *accumulation border layers* are formed in the in the insulation in front of both semiconductive layers. With continued stressing, a gradual equalization occurs due to the concentration gradient created between the high electron concentration in the border zones and the lower concentration in the remaining volume of insulation, as electrons penetrate into deeper levels of the dielectric.

  The consequence of this excess negative charge in the insulation layer is an increase in the potential gradient and thus in the electrical stress at the instantaneously positive pole while, at the same time, the field weakens at the negative pole (see Figure 3.3). In the limiting condition, the space charge field at the instantaneous cathode reaches the exact value of the geometrically impressed field strength and the sum of the two converges towards zero; the electron injection ceases. Conversely, following change of polarity, the geometrically impressed field and the charge field are added; the resulting stress on the now positive electrode is practically doubled in the steady-state condition compared to the case without space charge.

Unlike the homogeneous solid insulation material discussed so far, the multi-layer dielectric of an impregnated paper cable represents at least a two-material system and, in the case of internal gas pressure cable, perhaps a three-material system with cellulose, im-

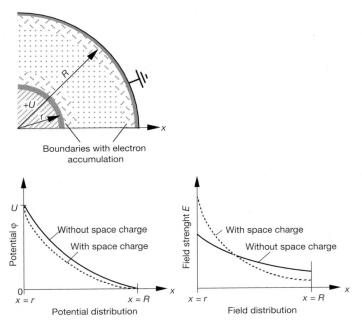

Figure 3.3
Electron enrichment and its effect on potential distribution and field pattern (schematic)

pregnating medium and possibly traces of pressurizing gas (see Section 2.2.3, Figure 2.3). These components form an interconnected network of unknown partial capacitances with cellulose and impregnating medium as the dielectric. This prevents an exact mathematical determination of the partial voltages and thus the actual field distribution within the insulation layer (see Figure 3.4). Nevertheless, the greatly simplified assumption of a series

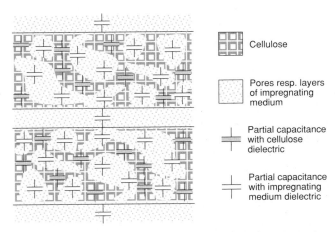

Figure 3.4   Structure of a paper cable dielectric and representation as a capacitance network

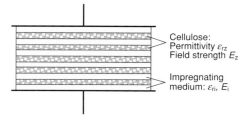

Cellulose:
Permittivity ε_{rz}
Field strength E_z

Impregnating
medium: ε_{ri}, E_i

Figure 3.5
Representation of a paper dielectric as a series circuit of layers of cellulose and impregnating medium

circuit of cellulose and impregnating medium layers as shown in Figure 3.5 is sufficient to estimate the most critical stressing.

In an arrangement of this type, the electrical field strength E of neighboring individual components behaves as inversely proportional to the relative dielectric constants ε_r of the respective insulation materials:

$$E_i / E_z = \varepsilon_{rz} / \varepsilon_{ri} \tag{3.1}$$

where i indicates the impregnating medium and z the cellulose layers. With values of ε_{rz} = 6.08 for pure cellulose and ε_{ri} = 2.2 for low-viscosity cable oil (valid at 50 Hz / 20 °C in each case [15], P. 160, 218), this results in a stress distribution of about E_i / E_z = 2.8. In other words, even in the absence of any impurities, geometric inhomogenities or space charges (which, in fact, play no part in paper-insulated cables), the impregnating medium zones are nearly three times as highly stressed as the cellulose.

3.1.2 Thermo-Mechanical Stresses

Cables are subjected to a series of thermo-mechanical stresses even before the commissioning of a new system and these stresses must be limited to a tolerable level for each type of cable to prevent any resultant damage. These stresses include:

- In the case of paper-insulated cables, the drying processes of the insulating paper and the impregnating medium which, for physical reasons, become quicker and more effective the higher the temperature used. Too high a temperature can lead to the beginning of decomposition of the paper or the vaporization of the more volatile fractions of the impregnating medium and can thus undesirably influence the chemical composition of the components (see Section 4.1 for details)

- In the case of polymer-insulated cables, the extrusion and, if applicable, the subsequent cross-linking processes are also usefully accelerated at increased temperatures. The viscosity of the melt is reduced, simplifying extrusion, and the subsequent cross-linking process takes place more quickly (see Section 4.2). Nevertheless, too high extrusion temperatures can cause scorching of the material and too high a thermal stress during cross-linking leads to the start of ageing of the material,

- In the case of all types of cable, the transport on drums and the cable laying, both of which are associated with substantial bending stresses. To prevent early mechanical damage in these cases, a lower limit is specified for the permissible bending radius of the cable depending on the type of cable, the conductor diameter and overall diameter, and the bending ability must be proven during type testing by repeated unrolling and re-rolling of suitable test drums (see Section 8.2).

Table 3.2
Permissible conductor temperatures for high and extra high voltage cables

Dielectric	Impr. paper	LDPE	HDPE	XLPE	EPR
Operating temperature in °C	85/90	70	80	90	90
Short-circuit temperature in °C	160/180	150	180	250	250

Permissible operating temperatures

Following the one-off stresses during manufacture and laying, cables are subjected to continual thermal stresses during operation according to load that build up additional mechanical stresses if a definite material-related upper limit is exceeded, above all accelerated *thermal ageing*. In the case of thermoplastic materials, this can lead to softening of the insulation layer and thus even to eccentric displacement (sagging) of the conductor. To prevent risks of this kind, the maximum permissible operating temperature for the particular type of cable is specified clearly in the applicable specifications. There is a continuous falling temperature gradient under load from the conductor via the insulation to the outer surface of the cable and, according to soil conditions, this continues in the surrounding ground. The specified maximum permissible continuous operating temperatures thus always refer to the conductor but the value is determined by the type of dielectric. In addition, many considerably higher transient over-temperatures occur in many cases, and the cable must be able to withstand these in the case of a short-circuit or earth fault for the duration of the so-called *rated short-circuit time* – usually 1 sec – without causing permanent damage to the individual components of the cable

Table 3.2 provides a summary of the limit temperatures of the most important types of high and extra high voltage cables. This shows that the cross-linked XLPE and EPR materials have high thermal reserves, even in the case of a short-circuit, one reason why cross-linked polyethylene is the most widely used cable dielectric world-wide in the high voltage range.

Thermal stabilization

All cables that can withstand the operating temperatures shown in Table 3.2 are, in principle, thermally stable. However, for historical reasons, the term thermal stabilization is generally only used for versions with impregnated paper dielectric. This is because, in the beginnings of cable technology, the heating unavoidable when transmitting high currents, in conjunction with high voltages, presented a problem that was not overcome for several decades (see Section 2.1).

The expansion of the impregnated paper insulation under the influence of heat causes bulging of the surrounding sheath (usually lead) and this bulging remains in place after subsequent cooling unless suitable preventive measures are employed. In the now increased volume under the sheath, the available amount of impregnating medium is insufficient to fill all the pores in the cellulose and the gaps between the individual paper layers as shown in Figure 3.4 (Section 3.1.1) without the formation of voids. This results in the creation of bubbles where there is no impregnating medium (*pinholes*) in which

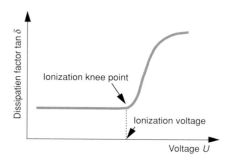

Figure 3.6
Ionisation knee point in loss factor curve of paper-insulated cables

ionization and pre-discharges occur above a particular electrical stress. The effect of these is to cause long-term damage to both the surrounding impregnating medium and the cellulose until, eventually, complete breakdown occurs (*erosion breakdown*). An accompanying effect of this breakdown mechanism is a pronounced rise in the loss factor depending on the voltage due to ionization losses beyond what is referred to as the *ionization knee-point* (see Figure 3.6).

Thermal stabilization has the effect of ensuring that no pinholes, in which stable discharges can be initiated by operating stresses or unavoidable network overvoltages, arise below the permissible temperatures in accordance with Table 3.2. Thermally stable cables are thus characterized in that no ionization knee-point occurs in the loss factor until clearly above the operating field strength.

The technologies employed in paper-insulated high and extra high voltage cables to prevent ionisable pinholes under changing thermal stresses can be divided into three categories (see Table 3.3):

Table 3.3 Thermal stabilization measures for paper-insulated cables

Measure	Effect	Application
Use of low-viscosity impregnating medium at medium positive pressure 1...5 bar, with expansion tanks connected by hollow channel in conductor	Impregnating medium flows into expansion tank on heating. On cooling, it flows back into the insulation and prevents the formation of pinholes	Low-presure oil-filled cables up to 500 kV
Use of Non Draining Compounds and gas-permeable core covering. Operation under 15 bar gas or oil pressure in steel pipe	Pressurized medium penetrates voids in the insulation and prevents ionization (according to Paschen's law)	Internal gas pressure cables up to 275 kV (110 kV in Germany), in Japan high-pressure oil filled cable up to 500 kV
Use of high-viscosity impregnating medium and gas-tight core covering (lead sheath), operation under 15 bar gas pressure in steel pipe	Insulation including lead sheath expands under heating. On cooling, it is compressed again without pinholes.	External gas pressure cables up to 275 kV (110 kV in Germany)

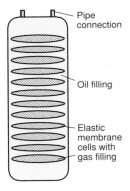

Pipe connection

Oil filling

Elastic membrane cells with gas filling

Figure 3.7
Expansion tank with
gas-filled membrane cells for
low-pressure oil-filled cables

- Use of low-viscosity impregnating media (cable oils) under moderate positive pressure (2 to 6 bar abs.) that flow on heating into external expansion tanks having gas or air-filled elastic membrane cells as shown in Figure 3.7. On cooling, the oil is automatically forced back into the dielectric. Flow channels of sufficient cross-section act as connections between the insulation and the equalization vessel. These usually take the form of a hollow channel in the center of the conductor (Figure 2.4). The insulation remains free of pinholes under all operating conditions. Application: low-pressure oil-filled cables up to 500 kV.

- Use of *non-draining compounds* as impregnating medium and filling any pinholes that exist or arise by using nitrogen N_2 under high pressure (15 bar) to surround the three phases in a common steel tube, the three phase cables being permeable to the gas. In accordance with Paschen's law, the pressurized gas causes the breakdown value of the pinholes to be raised to such a high field strength that no ionisation processes occur. Application: internal gas pressure cables up to 110 kV in Germany and up to 275 kV in Great Britain and USA. Low-viscosity oil, also at a pressure of 15 bar, can be used instead of N_2 as an "ionisation brake"; this design, known as high-pressure oil-filled cable or *Oilostatic* cable is used in the USA at up to 345 kV and in Japan at up to 500 kV.

- The use of a high-viscosity medium to impregnate the individual cores, which are enclosed in a gastight lead sheath; operation of the three phases of a system in a common steel pipe under an external gas pressure (N_2) of about 15 bar. The volume of the insulation that expands under the effect of heating is re-compressed on cooling without the formation of pinholes; a slight ovality of the core cross-section, achieved by the use of oval conductors as shown in Figure 2.4, enhances the so-called membrane effect of the lead sheath. This principle is a characteristic of external gas pressure cables, which are normally used in Germany only up to 110 kV, but in other countries (GB, USA) up to 275 kV.

The function of this last-described type of thermal stabilization and its effect on the loss factor is shown again schematically in Figure 3.8. The behavior of the insulation without external gas pressure as shown here would occur in normal medium voltage mass-impregnated paper cables if they were subjected to the same thermo-electrical stresses as high voltage cables. With a configuration largely identical to external gas pressure cables (but without the use of pressure) ionisation processes in mass-impregnated paper cables can only be excluded by drastic reduction of the operational stresses: max. 65 °C con-

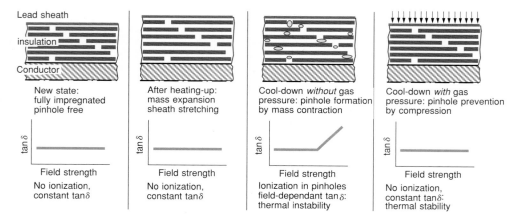

Figure 3.8   Principle of thermal stabilization using external gas pressure

ductor temperature and 4 kV/mm as compared with 85 °C and 12 kV/mm in the case of thermally stable cables.

Provided the dielectric is carefully manufactured (i.e. largely without impurities) and has permanently bonded semiconductive layers (see Section 4.2), thermal stability is provided at the design stage and the application of pressure is not necessary in the case of polymer-insulated cables. To prove the absence of pinholes, considerably more sensitive voltage-dependent partial discharge testing is used in this case rather than loss factor measurement (Section 8.3).

3.2 Dielectric Strength

The term "strength" derived from mechanics and used to describe the capacity of insulation or of an insulating material to withstand electrical stress does not in any way refer to a constant value. In fact, the dielectric strength of a material is considerably influenced by numerous parameters in the same way as the mechanical strength: temperature, form and frequency of voltage, field distribution, size of the stressed volume, duration of stress and many others that are discussed in detail elsewhere. If the dielectric strength of a cable insulation specified under definite conditions is exceeded, discharge processes always occur, and these can be divided into two categories; partial discharges and complete breakdown.

3.2.1 Partial Discharges

The occurrence of partial discharges (PD) within a dielectric – so-called *internal PD* – means that either the electrical field or the dielectric strength or both are distributed in a highly inhomogeneous manner. Figure 3.9 shows two classical examples: a pointed protrusion in a semiconductive layer that influences the field distribution on its own and a gas-filled cavity in the dielectric that disturbs both the field pattern and the distribution of the dielectric strength. On the other hand, the inhomogenity caused by the cylindrical geometry of the cable is not sufficient in its own to cause partial discharges.

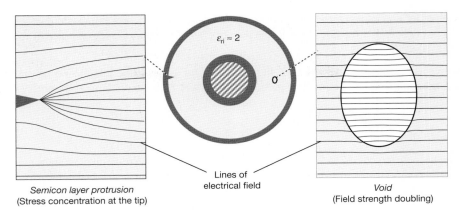

$\varepsilon_{ri} \approx 2$

Lines of electrical field

Semicon layer protrusion
(Stress concentration at the tip)

Void
(Field strength doubling)

Figure 3.9   Defects in solid dielectrics as origins of partial discharges

In the case of the pointed protrusion in the semiconductive layer, the danger exists of electrically overloading the dielectric in the area of the greatest concentration of field lines in the direct vicinity of the defect. As a result of this, the material is locally separated and an initial hollow channel forms, a few μm long. From then on, stable gas discharges take place within this channel that cause further erosion of material in the form of continuing *channel growth,* and these discharges can be detected externally as high-frequency partial discharge pulses using suitable equipment.

On the other hand, the existence of a gas-filled cavity permits the onset of partial discharges without previous damage to the material having occurred. This is because the field strength inside the void ($\varepsilon_r = 1$) exceeds the stress in the surrounding dielectric by a factor of nearly ε_{ri}, the permittivity of the insulation material. At the same time, gases usually have a considerably lower dielectric strength than solid, liquid or impregnated dielectrics. As a result, discharges occur in the void above a definite voltage that can be measured externally and can lead to a gradual erosion of the surrounding material.

The degree of risk to a high voltage cable due to internal partial discharges depends, apart from the cause and intensity of the partial discharge, above all on the type of dielectric – impregnated paper or polymer – and its composition.

Impregnated paper dielectric

The many years of operating experience show this type of cable insulation to be largely insensitive to partial discharges. Two facts are principally responsible for this:

- Any voids are relatively quickly filled again by the presence of a liquid impregnating medium, extinguishing any partial discharges. These conditions exist in low-pressure oil-filled and external gas pressure cables. This possibility does not exist when non-liquid impregnation materials are used as in internal gas pressure cables; instead, gas under external high pressure penetrates the voids and increases the ignition voltage of partial discharges to such a level that no discharges occur under operating conditions (these properties have already been described as characteristics of thermally stable dielectrics).

- Even if local discharge processes still occur for a limited time, any associated damage process is halted at the latest due to the barrier effect of the next layer of paper of the multi-layer dielectric.

Taking this knowledge into account, the standards for paper cable contain no recommendations for the measurement of partial discharges at the manufacturer's premises or on site (see Section 8).

Extruded dielectric

Completely different conditions exist in extruded polymer dielectrics: they contain neither liquid components to fill the voids nor barriers to halt the growth of channels. As a result, internal partial discharges in this type of material usually lead inexorably to complete breakdown, known as *erosion breakdown*. Differences between the insulation materials in question for high voltage cables (PE/XLPE or EPR) occur in terms of the time between the onset of the first partial discharges and complete breakdown, PE and XLPE being especially sensitive to the effects of partial discharges.

One of the reasons for this is that the erosion of this material leads mainly to the production of gaseous, non-conductive decomposition products such as hydrogen, methane and low-molecular hydrocarbons [16]. These are unable to form conductive deposits on the channel walls or the surfaces of the voids and thus to electrically de-stress the discharge space, at least for a period.

EPR is in fact more resistant to the effect of partial discharges but, even with this material, the time span between the beginning of the first stable discharges and the resultant erosion breakdown is far too small in relation to the expected life of the cable to allow measurable partial discharges to be tolerated.

As a result, the requirement for total freedom from partial discharges applies as much to EPR as to PE/XLPE cables in terms of the minimum intensities that can be detected reliably by test technology (Section 8).

Electrical stabilization

The knowledge outlined above that internal partial discharges in polymer-insulated cables, once started, always lead to failure led to intensive development activities in the 1960s and 1970s with the objective of preventing the onset of partial discharges in the first place. The starting point for these electrical stabilization measures, intended mainly for thermoplastic polyethylene, was the idea of electrically neutralizing the unavoidable inhomogenities in the dielectric by the use of suitable additives, known as *voltage stabilizers* [17].

To implement this effect, the additives must have a higher dielectric constant and/or conductivity relative to the surrounding dielectric with alternating voltages and they must be deliberately concentrated in the immediate vicinity of the impurities, i.e. unevenly distributed within the insulation. Figure 3.10 shows schematically the field reduction around a semiconductive layer protrusion where a stabilization layer with increased dielectric constant has been introduced. On the other hand, an equivalent enrichment of the entire volume of insulating material with such an additive would have no effect on the field distribution.

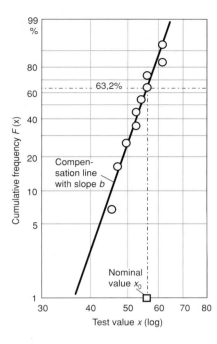

Figure 3.17
Evaluation of test results
as simple distribution
in the Weibull diagram

effect is a purely statistical phenomenon. A *statistical volumetric effect* of this kind can generally be assumed in the case of correctly manufactured polymer-insulated cables, i.e. the probability of the existence of a strength-reducing weak point increases as the insulation volume increases – no matter whether this is due to an increase of insulation thickness or a greater cable length. With this precondition, the breakdown field strength $E_D(V_2)$ of the volume V_2 can be calculated from the known dielectric strength $E_D(V_1)$ of the insulation volume V_1 by using the Weibull slope b from Equation 3.7 [28]:

$$E_D(V_2) / E_D(V_1) = (V_1 / V_2)^{1/b} \qquad (3.9)$$

- *Probability transformation*

Extremely high reliability is expected from the equipment used in power supply systems. The practical user is therefore less interested in the arithmetical average value or the 63.2% nominal value of the breakdown field strength but rather in an electrical stress that corresponds to a much lower probability of failure. The appropriate transformation is achieved by using Equation 3.8 where, in this case, the breakdown field strength E_D is used in place of the general value x and the probabilities p_i replace the cumulative frequency $F(x)$:

$$\frac{E_D(p_1)}{E_D(p_2)} = \exp\left[\frac{\ln[-\ln(1-p_1)]-\ln[-\ln(1-p_2)]}{b}\right] \qquad (3.10)$$

where p_1 and p_2 are the failure probabilities being considered.

In addition to these two transformation equations derived from the Weibull statistics, another conversion formula is needed to determine the insulation thickness of cables with extruded insulation; this formula determines the breakdown field strength $E_D(t_2)$ after t_2 from a known dielectric strength $E_D(t_1)$ after the stressing time t_1 :

- *Time transformation*

  This results directly from the empirical life law, Equation 3.4:

  $$E_D(t_1) / E_D(t_2) = (t_2 / t_1)^{1/N} \qquad (3.11)$$

  where N is the constant predetermined life exponent.

Calculation of insulation thickness

Because of their pronounced life characteristic, the dimensioning of cables with polymer dielectric is unlike that for paper-insulated cables (Section 3.3.1) in that all types of stress have to be taken into account under test and operating conditions in both short-term and long-term areas as there are no binding standards available.

One possibility is the use of a calculation method originally published by Japanese manufacturers, known as the Furukawa method [29], which was later modified and extended by Siemens [21, 22, 30]. With this method, the insulation thickness w is calculated as a quotient from the so-called *design voltages* U_d and *design field strengths* E_d for impulse and long-term AC voltage stresses, the larger of the two values determined being taken as the provisional insulation thickness:

$$w = U_d / E_d \qquad (3.12)$$

The design voltages have a direct relationship to the rated voltages and the impulse levels standardized for these, while the design field strength basically represents manufacturer-specific product properties and must therefore be supported by appropriate test results. These field strengths are, in principle, of a safe level that will not lead to any damage (*withstand field strengths*).

The following applies for AC voltage (*method A*):

$$w = U_{dac} / E_{dac} = U_0 \cdot k_T \cdot k_{\ddot{u}} \cdot k_t / E_{dac} \qquad (3.13)$$

where

E_{dac} = design AC field strength
U_0 = conductor-to-screen operating voltage
k_T = temperature factor, depending on insulation material (XLPE: $k_T = 1{,}25$)
$k_{\ddot{u}}$ = voltage magnification factor, $k_{\ddot{u}} = 1.15$
k_t = ageing factor, takes account of the loss of dielectric strength during the course of the nominal service life of the cable compared to 1h stressing time

Factor k_t is determined from two quantities, the calculated nominal service life t_N and the life exponent N and is calculated in a similar way to Equation 3.11:

$$k_t = (t_N / 1\text{h})^{1/N} \qquad (3.14)$$

The value of t_N is usually 30 to 40 years; on the other hand, the life exponent depends on the cable dielectric and must be determined from long-term tests that are carried out as realistically as possible. Published guide values for XLPE extra high-voltage cables are, in any case, substantially greater than $N = 10$; more recent tests show $N = 15$ to 20 to be realistic [10, 21].

The relationship for impulse voltage, corresponding to eq. (3.13) (*method B*) is:

$$w = U_{ds} / E_{ds} = U_{rB} \cdot k_T \cdot k_f \cdot k_s / E_{ds} \tag{3.15}$$

where E_{ds} = design impulse field strength
 U_{rB} = rated withstand lightning impulse voltage
 k_T = temperature factor as in eq. 3.13
 k_f = repetition factor, $k_f = 1,1$
 k_s = safety factor $k_s = 1,1$

The modified method derived from the classic Furukawa method takes account of the knowledge that impulse voltage breakdowns in technically fault-free polymer cable insulation occur first at the inner semiconductive layer where the electrical stress is at its highest [29]. The provisional insulation thickness calculated using methods A and B must therefore be checked, especially for small conductor cross-sections (greater field inhomogeneity), under impulse stress using the following *method C*:

$$U_{ds} = w \cdot E_{dsmax} \cdot \eta \tag{3.16}$$

where $\eta = r \cdot \ln \left[(r + w) / r \right] / w$ $\qquad$ (3.17)

as the *Schwaiger factor* for cylindrical arrangements

and U_{ds} = design impulse voltage from Equation 3.15
 E_{dsmax} = design impulse field strength at the inner semiconductive layer
 r = radius of the inner semiconductive layer.

The combination of Equations 3.16 and 3.17 then results in

$$w = r \cdot \left[\exp(U_{ds} / r \cdot E_{dsmax}) - 1 \right] \tag{3.18}$$

As a rule, taking into account additionally the field concentration at the inner semiconductive layer according to method C results in the insulation having to be made thicker for very small conductor cross-sections compared to the thickness calculated for homogeneous field distribution (methods A and B). Figure 3.18 illustrates this with an example of the insulation thickness for 500 kV XLPE cable [21].

In addition, some national specifications, e.g French specification C 33-253, require theoretical/experimental evidence from the manufacturers that cable systems manufactured by them do not exceed specified failure probability. Details of this are described in [21, 22] and other publications.

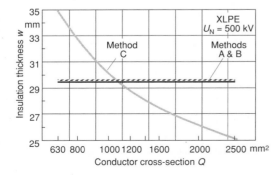

Figure 3.18
Insulation thickness for 500 kV XLPE-cable (in accordance with [21])

3.4 Losses in the Cable

The losses that occur in a cable system under operating conditions are not only worth calculating more accurately from the economic point of view, but they can also cause local overheating, accelerated ageing or even premature failure of the insulation. The total losses can be divided into two types:

- Losses that only occur if current is flowing in the cable system (*current-related losses*) and

- Losses that are produced solely by the effect of the electrical field on the insulation (*voltage-dependent losses*).

3.4.1 Current-Dependent Losses

These occur not only in the current-carrying conductor itself (*conductor losses*) but also as so-called *additional losses* in the other metallic elements of a cable system where eddy and circulating currents are induced under the effect of the magnetic field of the current-carrying conductor.

Conductor losses

It has already been mentioned in Section 2.3.1 (Figure 2.5) that the conductor resistance responsible for the ohmic losses can considerably exceed the material-specific DC value as a result of less than optimum utilization of the cross-section in AC operation due to skin and proximity effects. Because of the intrinsic magnetic field created by the conductor current, the skin effect, especially, causes an increase in the current density at the outer area of the conductor while the center carries less load. Since, for thermal reasons (max. conductor temperature), specified maximum values of current density may not be exceeded, even in the outer zones, the unequal utilization of the cross-section results inevitably in reduced current carrying capacity of the conductor (Figure 3.19).

The International Electrotechnical Commission (IEC) looked into the subject of the quantitative estimation of increased resistance in the 1960s and summarized the methods

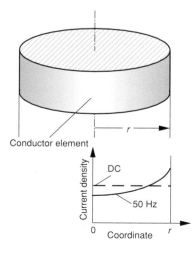

Figure 3.19
Reduction of current carrying capacity of the conductor due to current displacement in the outer area

of approximation in the publication IEC 287 [5]. The starting equation is the theoretical statement

$$R_\sim / R_= = (1 + y_s + y_p) \tag{3.19}$$

where y_s and y_p are the skin and proximity coefficients that represent the current displacement caused by each effect. The influence of the proximity effect (coefficient y_p) as a result of the magnetic field created by the adjacent cables of the three-phase system can usually be neglected for single core cables under realistic laying conditions. Experimental tests in [6] have shown that practically no increase in resistance due to the proximity effect can be detected where the spacing of the conductor centers is greater than 30 cm. This component of current displacement only needs to be taken into account in the case of gas pressure cables with all three phases in a common steel pipe or single-core cables laid in threefoil formation. Nevertheless, even this only has a significant effect with conductor cross-section above 1000 mm², which are uncommon in gas pressure cables, at least in Germany.

In practice, increase in resistance is thus caused almost entirely by the skin effect. The approximation formulae 3.20 and 3.21 can be used to calculate the associated coefficient y_s:

$$y_s = x_s^4 / (192 + 0.8\ x_s^4) \tag{3.20}$$

where $\quad x_s = \sqrt{2\ \mu f k_s / R_=}$ \hfill (3.21)

and $\quad \mu$ = permeability
$\quad\ \ f$ = frequency
$\quad\ \ k_s$ = geometric factor.

k_s represents a dimension for the conductor geometry and filling factor. Its value is therefore determined by the conductor design, compression and wire pretreatment and is in any case ≤ 1 ($k_s = 1$ applies to circular solid conductors without hollow center). For stranded conductors, k_s can only be determined experimentally. Favorable designs such as segmental or Milliken conductors have small k_s parameters below 0.5, resulting in the comparatively favorable resistance behavior of these types of conductor. However, a really significant reduction of current displacement by the use of segmental conductors only occurs in conductors of large cross-section in conjunction with additional individual insulation of the wires as shown in Figure 3.20. Without this, the segmented conductor only acts to reduce the proximity effect and therefore has properties hardly any bet-

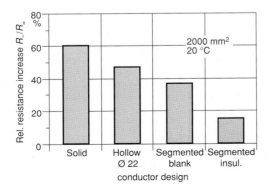

Figure 3.20
Increase in resistance of various types of copper conductor due to the skin effect at 50 Hz

ter than those of a profiled wire conductor of the same size with a 22 mm hollow channel, as used for low-pressure oil-filled cables [6].

In absolute terms, the conductor losses, which increase as the square of the current, are of the order of a few tens of kW/km in high-voltage cable systems operating at 50/60 Hz and rated load. This is of minor importance from the energy economy point of view with transmission powers of several 100 MVA per system. Nevertheless, problems result from the fact that these losses – together with the further losses in the metallic sheath and in the insulation as discussed below – must be dissipated through the soil into the atmosphere and can thus lead to drying of the soil, causing dangerous build-up of heat (see Section 3.5).

Additional losses

This term includes all current-dependent losses that arise in the metallic components of a cable system *external to* the conductor. These include *induction current losses, eddy current losses* and *magnetic reversal losses*. These, also, can be formally added to the conductor resistance in a similar manner to Equation 3.19 as additional parameters λ_i, the index i referring to the source of the additional loss. λ_3, for example, indicates the loss in the cable sheath, λ_6 indicates losses in the metallic armouring. There are also empirical approximations for the calculation of the λ_i parameters under particular system configurations in IEC Publication 287 [5].

The most important additional losses are those caused by axial induction currents in the metallic sheath. As shown in Figure 3.21, this arises due to the induction of a longitudinal voltage U_i in the cable sheath of one phase due to the magnetic field of the current-carrying conductors of the adjacent phases. For a particular frequency and plant configuration, the value of the *induction voltage* is directly proportional to the value of the causative currents and the length of the cable run in question. Under unfavorable conditions – i.e. especially in the case of a short circuit – values of a few kV can occur.

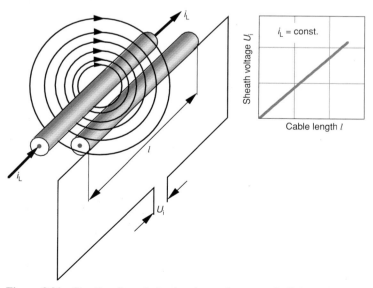

Figure 3.21 Sheath voltage induction due to the magnetic field of an adjacent phase

To avoid any resulting danger to persons or equipment, the cable sheaths of extensive systems of this type are generally solidly earthed at both ends. This, however, creates a closed loop in which equalization currents flow under the effect of the induction voltage and, in conjunction with the finite sheath and/or screen resistance, can cause considerable current heating losses. Since the induced sheath voltage increases in proportion to the conductor current independent of the conductor design, the sheath losses can rise to levels of the same order as the conductor losses without suitable preventive measures, and can even exceed these significantly at high rated currents and correspondingly large conductor cross-sections.

Compared to these induction current losses, the other additional losses due to eddy currents and magnetic reversal effects in cable sheaths, armoring and other metallic construction elements are of secondary importance. However, as they are additional heat sources, they may adversely affect the thermal stability of the cable and must therefore be estimated for temperature calculations based on IEC publication 287 [5] and, in borderline cases, reduced by suitable measures.

Reducing the current-dependent losses

The most important measures for reducing the AC losses in the conductor are the use of a suitable design of conductor to reduce the skin effect (segmental or Milliken conductor, if possible with individual wire insulation, see Section 2.3.1, Figure 2.5) and, in the case of single core cables, laying the cables at a minimum separation of 30 cm to eliminate the proximity effect. In conjunction with low-pressure oil-filled cables which, in any case, need a hollow channel in the conductor to supply impregnation medium, enlargement of the hollow channel diameter offers an effective means of reducing the current displacement. In this way, the utilization of the material is transferred from the less-used center of the conductor to the highly-loaded outer area. Additionally, it has been proved that arranging the phases in an equilateral threefoil configuration results in better utilization of the cross-section than laying them side by side, no matter what the design of conductor.

The frequently equally important additional losses in the cable sheath, screen and other metallic system components can be reduced by various measures that usually affect the entire system concept, including:

- *Use of non-magnetic steel* for armoring to prevent magnetic reversal losses

- *Grounding* of screens or metal sheaths *at one end only* to avoid providing a closed loop for the induction currents. This very effective measure is, however, limited to very short runs (e.g. from switchgear to transformer), as the sheath voltage increases in proportion to the length of the parallel-lying phases as shown in Figure 3.21 and has a dangerously high value at the open end

- *Cross bonding* of the cable sheaths to largely compensate the induction voltages so that sheath current and consequent losses are minimized in spite of the entire run being grounded at both ends. As shown in Figure 3.22, the cable run is divided into three or a multiple of three sections of as nearly as possible equal length and solidly grounded at the outer ends. A cyclic change of the sheath connections at the two inner joints ensures that the induction loops are exposed to the magnetic field of the conductor in all three possible positions respectively and over the same length. Thus, in the ideal case, the longitudinal voltages between the two ground connections add (vectorially) to zero and cir-

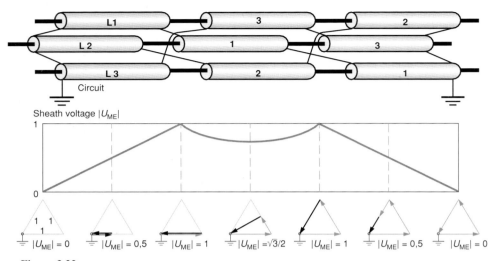

Figure 3.22
Complete compensation of induced sheath voltages by cross bonding with cables laid in threefoil configuration

culating currents no longer occur. Nevertheless, this ideal condition is linked to at least two further conditions in addition to the exactly equal lengths of the sections between the crossover points and, in practice, these conditions are not completely satisfied [31]:

– The mutual inductance between the cable sheaths and the neighbouring inducing conductors must be the same for all three phases at every point of the cable run, a condition most closely approximated by a phase configuration in an equilateral triangle and

– The conductor currents must be of identical value and must have a phase displacement of exactly 120° to each other.

These conditions are the basis of the schematic representation of the graphs of the additive sheath voltages to ground shown in Figure 3.22.

Although, in practice, complete compensation of this kind cannot be achieved (for example, identical mutual induction basically cannot be achieved in the preferred flat configuration for laying extra high-voltage cable systems), cross bonding is nevertheless the most effective measure for the reduction of current-dependent sheath losses. As a result, it is nowadays used in all high-power cable systems. Cross bonding also requires the use of additional special *isolating joints* to ensure a sufficient isolation distance between the sheath voltages at either side of the joint (see Section 6.3.3).

Table 3.5 provides a keyword summary of the measures described above for the reduction of current-dependent losses.

3.4.2 Voltage-Dependent Losses

Voltage-dependent losses arise exclusively in the dielectric of a cable and, for this reason, are often known as *dielectric losses*. The determining factors for their total value P_{diel} per cable system under operating conditions are the voltage U_0 across the insulation layer, the capacitance C per phase and the dielectric loss factor $\tan\delta$:

Table 3.5 Measures for reducing the current-dependent losses

Measure	Effect	Practical application
Use of Milliken conductors with insulated segments	Reduction of proximity effect also reduces the skin effect at low compaction	XLPE cables above 1000 mm² (Cu)
Use of Milliken conductors with insulated single wires	Almost complete prevention of current displacement in conductor	Extreme cross-sections ≥ approx. 2000 mm²
Use of hollow conductors with enlarged channel diameter	Reduction of the skin effect	LP oil-filled cables from approx. 2000 mm²
Phase separation >30 cm	Minimization of proximity effect	Cross-sections ≥ 1000 mm²
Threefoil phase arrangement	Reduction of proximity effect and sheath voltage induction	Cross-sections ≥ 1000 mm²
Use of non-magnetic armoring materials	Prevention of magnetic reversal losses	Cross-sections ≥ 1000 mm², where armoring is necessary
Sheath grounding at one end only	Prevention of induced sheath currents	Short systems up to about 500 m
Cross bonding of cable sheaths	Compensation of sheath voltages	Cross-sections ≥ 1000 mm²

$$P_{\text{diel}} = 3\, U_0^2\, \omega\, C \cdot \tan\delta \qquad\qquad (3.22)$$

where $\omega = 2\,\pi\,f$ as the angular frequency at a network frequency of f.

Whereas the dissipation factor in Equation 3.22 represents a value that, although influenced by various parameters, finally depends on the material, the capacitance C depends principally on the dimensions of the cable system, especially its length. Using the *volume-specific power loss* P'_{diel} avoids this arbitrary parameter:

$$P'_{\text{diel}} = E_0^2\, \omega\, \varepsilon_0\, \varepsilon_r \cdot \tan\delta \qquad\qquad (3.23)$$

where E_0 = operating field strength
  ε_0 = permittivity of a vacuum ($\varepsilon_0 = 8.855 \cdot 10^{-12}$ As/Vm)
  ε_r = permittivity of the insulation material.

From the above, the value of power loss per volume element produced in the insulation is determined, above all, by the product of permittivity and dissipation factor $\varepsilon_r \cdot \tan\delta$ in addition to the field strength. This product is therefore also known as the *dielectric loss coefficient*, a dimensionless parameter that is one of the most important material properties of the cable dielectric. It follows from this that the relatively high operating field strengths of high-voltage and extra high-voltage cables only permit the use of materials with extremely low loss coefficient.

Dissipation factor and loss coefficient of different materials

In the case of macroscopically homogeneous high-voltage insulation materials, the dissipation factor, permittivity constant and loss coefficient are determined by the composition of the materials and only show relatively minor deviations when subjected to the thermal and electrical stresses that occur during operation of the cable and under test

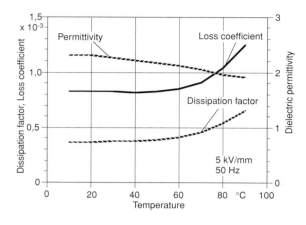

Figure 3.23
Dielectric properties of XLPE as
a function of temperature

conditions. As *non-polar* insulation materials, PE and XLPE (see Section 4.2.1), in particular, exhibit almost ideal values in this respect. As an example, Figure 3.23 shows the temperature curves of the three parameters for an XLPE cable insulation including the semiconductive layers.

On the other hand, impregnated paper dielectrics are not homogeneous materials but a mixture of cellulose and impregnating medium, each having very different dielectric properties (Table 3.6). As the combination of cellulose and impregnating medium is of a more complex nature than simple parallel or series circuits (see Figure 3.4), relatively complex formulae must be used to calculate the resultant properties of the mixed dielectric:

$$\varepsilon_{rk}^{\alpha} = v_z \cdot \varepsilon_{rz}^{\alpha} + v_i \cdot \varepsilon_{ri}^{\alpha} \tag{3.24}$$

and $\quad \tan\delta_k = \dfrac{v_z \, \varepsilon_{rz}^{\alpha} \, \tan\delta_z + v_i \, \varepsilon_{ri}^{\alpha} \, \tan\delta_i}{v_z \, \varepsilon_{rz}^{\alpha} + v_i \, \varepsilon_{ri}^{\alpha}} \tag{3.25}$

where

$\varepsilon_{rk}, \tan\delta_k$	= resultant permittivity and loss factor of the mixed dielectric
ε_{rz}	= permittivity of cellulose; $\varepsilon_{rz} = 6.08$
$\tan\delta_z$	= dissipation factor of cellulose; $\tan\delta_z = 4.3 \cdot 10^{-3}$
ε_{ri}	= permittivity of impregnation medium; $\varepsilon_{ri} = 2.2$ for oil
$\tan\delta_i$	= disipation factor of impregnation material; $\tan\delta_i = 0.2 \cdot 10^{-3}$
v_z	= rel. volumetric proportion of cellulose; $v_z = 0.5\text{–}0.7$[1]
v_i	= rel. vol. proportion of impregnating medium $v_i = 0.3\text{–}0.5$[1]
α	= mixed form exponent; $\alpha \approx -0.5$

Table 3.6
Properties of cellulose and cable oil at 50 Hz/20 °C

Property	Cellulose	Oil
$\tan\delta \cdot 10^3$	4.3	0.2
ε_r	6.08	2.2

---

[1] Volumetric proportions apply taking into account the impregnating medium gap between the paper layers

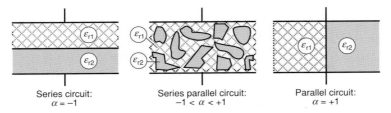

Series circuit:
$\alpha = -1$

Series parallel circuit:
$-1 < \alpha < +1$

Parallel circuit:
$\alpha = +1$

Figure 3.24   Characteristic areas of the mixed form exponent α

The mixed form exponent α describes the type of capacitive link between the individual components of a mixed dielectric. It can have values between -1 (pure series circuit) and $+1$ (pure parallel circuit) (see Figure 3.24). α is thus determined by the structure of the cellulose, a value of $\alpha = -0.5$ being the closest approximation to the practical relationships for high-voltage cable papers with raw densities between approx. 0.8 and 1.2 g/cm³ (pure cellulose: 1.53 g/cm³), i.e. a predominantly capacitive link in the form of a series circuit in the case of the paper structure.

It follows from formulae 3.24 and 3.25 and the values of the properties of the individual components (Table 3.6) that increasing the proportion of impregnating medium in the cable dielectric can theoretically have a positive influence on the loss coefficient $\varepsilon_r \cdot \tan\delta$, the decisive parameter determining the losses, since good impregnating medium always has considerably lower dissipation factors and permittivity values than cellulose. Figure 3.25 shows as an example the relationship between the cellulose component v_z of an oil-impregnated paper dielectric and the resultant permittivity. There are, however, two important reasons against such a measure:

- The fact, already highlighted in Section 3.1.1 (Equation 3.1), that the field strength within the impregnation medium zone is nearly three times as high as in the cellulose and

- The low dielectric strength of the liquid in comparison with paper (see Section 4.1.2., Figure 4.7).

As a result, increasing the proportion of the impregnation medium would increase the probability of breakdown of the cable insulation to an unacceptable degree. In practice, paper dielectrics have dissipation factors of approx. 2 to 3 · 10⁻³ and permittivity values of about 3.5 with either low or high viscosity impregnating media; the resulting loss coefficient is ≤ 0.01.

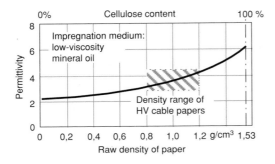

Figure 3.25
Relationship of the permittivity of oil-impregnated paper dielectric to the cellulose content

Table 3.7
Dissipation factor, permittivity and dielectric loss coefficient for various cable dielectrics at 50 Hz (guide values)

Dielectric	$\tan\delta \cdot 10^3$	ε_r	$\varepsilon_r \cdot \tan\delta \cdot 10^3$
Impregnated paper	2...3	3.5	7...10
Polyethylene	0.2...0.4	2.2...2.3	0.4...0.9
XLPE	0.3...0.5	2.3...2.4	0.7...1.2
EPR	1.8...3	2.6...2.7	4.7...8.1
Impreg. PPL	0.5...0.6	2.6	1.3...1.6

A summary of the values of the most important dielectric properties for the dielectrics used in high-voltage and extra high-voltage cables is provided in Table 3.7. The values shown here increase slightly for measurements on complete cables, especially for low-loss PE and XLPE (see Figure 3.23), as polarization processes within the only finitely conductive stress-relieving layers are superimposed on the intrinsic conduction mechanism of the insulation material [11].

Possibilities for reducing the dielectric losses

It can be seen from Equation 3.22 that the dielectric power loss is determined by just three parameters: the operating voltage, the loss factor and the cable capacitance which, in turn, depend solely on the permittivity with given dimensions. The most important influence is that of the voltage, as the value of the power loss is proportional to its square (Figure 3.26). For this reason, only in extra high-voltage cables do these voltage-dependent losses reach an order of size that make it necessary, in certain cases, to take measures to reduce these losses.

While the dielectric losses in three-phase XPLE cable systems, even with $U_N = 500$ kV, are always of a non-critical order of considerably less than 10 kW/km because of the outstanding insulation material properties, the comparative vales of common paper-insulated cables, with losses of around 35 kW/km at 380 kV or even 50 to 55 kW/km at 500 kV, can cause considerable heat dissipation problems when laid normally in the ground (see next Section). This is even more applicable, as the voltage-dependent losses are added to current-dependent losses of at least equal amounts. In the case of 750 kV cables, the limits of natural cooling are reached with the dielectric losses alone and power transmission is only possible using forced cooling.

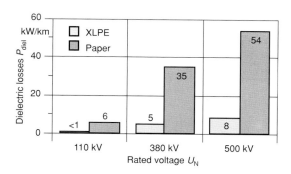

Figure 3.26
Dielectric losses of XLPE and paper-insulated cable systems

Unlike current-dependent losses, which can be effectively influenced using constructional and connection methods (Milliken conductor, cross bonding etc.), the options for
reducing the voltage-dependent losses for a specific dielectric are severely limited. Here,
increasing the volume of the impregnating medium at the expense of the high-loss cellulose, as previously suggested, can be disregarded from the beginning because of the high
field strengths that exist (for this reason, cable papers for extra high-voltages are characterized by a higher cellulose content compared with those for medium voltage cables.
Nevertheless, their loss factor can be reduced by using selected cellulose and special
preparation from about $2 \cdot 10^{-3}$, the value for good HV papers, to $1.5 \cdot 10^{-3}$).

Summarized, this means that a really significant reduction of the dielectric losses can
only be expected if the cellulose content of the impregnated dielectric is minimized and
replaced by a low-loss material with equally low permittivity *without* this leading to a
reduction in dielectric strength.

To this end, Japanese and UK-manufacturers promoted the development in the 1980s
of *low-loss papers*, i.e. papers that were either coated with polypropylene PP on one
side or applied together with separate PP foil [32]. Pure dodecyl benzene (DDB) has
proved to be a suitable low-viscosity impregnating medium for this kind of *PPL*
(*P*olypropylen *P*aper *L*aminate) cable, as it is chemically compatible with both paper
and PP under all operating conditions and its other properties are comparable with
high-grade mineral-oil-based cable oil. Because of its superior ageing behavior, DDB
is often also used for the impregnation of conventional paper-insulated cable, or at least
added by up to 20 % to mineral oil (see Section 4.1.1). The operating pressure and
maximum conductor temperature (85 °C) of PPL cables are the same as the comparable values for low-pressure oil-filled cable.

The dielectric properties that can be achieved using paper-polypropylene sandwich insulation are determined from the volumetric proportion of the two components in accordance with Equations 3.24 and 3.25. Since polypropylene has similarly favorable characteristic values to polyethylene, it is desirable to have as large as possible a proportion
of PP. In fact, for this reason, high-voltage dielectrics consisting entirely of impregnated
polymer foil were developed in the 1960s for use in instrument transformers, bushings
and capacitors [33]. However, a gaseous impregnating medium, the insulating gas *sulfur
hexafluoride SF$_6$*, was used, because liquids had proved to be problematic when used for
the impregnation of polymer foil. Nevertheless, this technology found just as little application in cables as the parallel development of liquid-impregnated polypropylene foil
with roughened surface, which was primarily intended for use in high-voltage power
capacitors.

In contrast, the design of PPL cables represents a compromise between the excellent
dielectric properties of pure polypropylene and the ability to impregnate the paper without problems. The volumetric proportion of PP to paper is between approximately 1:1
and 3:2 [34], the paper itself consisting only about 50 to 70 % of cellulose and the rest
being filled with impregnating medium. The dielectric properties achieved by this
method are shown in the last column of Table 3.7.

Nowadays, PPLP cables are available up to a rated voltage of 500 kV and have already
successfully withstood tests for the next higher voltage level of 750 kV [34]. The particularly important voltage-dependent losses in this case are more than 60 % lower than the
comparable values of traditional oil-impregnated paper cable.

Transmission voltage limits

In a similar manner to the capacitive charging current that limits the maximum length of three-phase cable systems (see Section 2.2.1, Figure 2.1), the maximum achievable transmission voltage for a given system configuration depends largely on the dielectric losses that rise as the square of the operating voltage. The heat created by these losses must be extracted from the cable system along with that created by the current-dependent losses. In *thermal equilibrium*, the heat extracted per unit length P'_{ab} is equal to the sum of the power losses P'_v in the same unit length:

$$P'_{ab} = \Sigma \ (P'_v) \tag{3.26}$$

where

$$\Sigma \ (P'_v) = P'_{vi} + P'_{vu} \tag{3.27}$$

and
$$\begin{aligned} P'_{vi} &= \text{current-dependent losses per unit length} \\ P'_{vu} &= \text{voltage-dependent losses per unit length.} \end{aligned}$$

With given soil properties and system configuration, P'_{ab} represents a finite constant power with a value of 50 to 60 W/m for a three-phase underground system without forced cooling. The sum of the power losses must not exceed this value and thus there is a relationship between the transmission voltage U, the maximum conductor current I_{max} and the maximum possible transmission power S_{max} as a combination of both values. The following is valid for a three-phase system:

$$S_{max} = U \cdot I_{max} \cdot \sqrt{3} \tag{3.28}$$

$$P'_{vi} = 3 \ I^2 \cdot R' \tag{3.29}$$

$$P'_{vu} = U^2 \ \omega \ C'_0 \cdot \varepsilon_r \cdot \tan\delta \tag{3.30}$$

with $C'_0 = (2 \ \pi \ \varepsilon_0) \ / \ln(D/d)$ $\hspace{3cm}$ (3.31)

Where:
$$\begin{aligned} R' &= \text{length-specific effective conductor resistance in } \Omega/\text{m} \\ C'_0 &= \text{length-specific vacuum capacitance per phase in F/m} \\ \omega &= \text{angular frequency } 2 \ \pi \ f \text{ with } f \text{ as network frequency in 1/s} \\ \varepsilon_r, \tan\delta &= \text{dielectric properties of the insulation} \\ \varepsilon_0 &= \text{permittivity of vacuum} \\ &\quad (8.855 \cdot 10^{-12} \text{ As/Vm}) \\ D, d &= \text{external and internal diameters of the insulation} \end{aligned}$$

The combination of this system of equations with the thermal equilibrium condition (Eq. 3.26) then provides a formula for the maximum transmittable power containing the line-to-line voltage U and the dielectric parameters ε_r and $\tan\delta$ as the most important influencing values:

$$S_{max} = \frac{U}{\sqrt{R'}} \cdot \sqrt{P'_{ab} - U^2 \omega C'_0 \cdot \varepsilon_r \cdot \tan\delta} \tag{3.32}$$

Knowing the type of conductor, its cross-section, the system configuration (and thus the effective conductor resistance R' per unit length), the thermal power dissipation and the insulation thickness for the particular operating voltage to allow the vacuum capacitance to be determined, it is possible to calculate the transmittable power for the cable system in question in relation to the voltage and the data for the various dielectrics. Figure 3.27 contains the results of suitable approximation calculations for systems with 1600 mm² Cu conductors of Milliken design (for XLPE and EPR) and as hollow conductor with

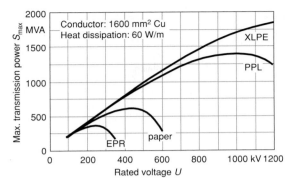

Figure 3.27
Transmitted power in relation to the voltage for cable systems with various dielectrics (explanations in the text)

profiled wires (for impregnated paper and PPL). The thermal power that can be dissipated (in the case of natural cooling) is in these cases about 60 W/m, the average effective conductor resistance is assumed as $1.46 \cdot 10^{-5}$ Ω/cm at 50 Hz and $85\ldots90\,°C$, referring to measurements made by Metra und Rühe [6], and the dielectric parameters ε_r and $\tan\delta$ are taken from Tables 3.6 and 3.7.

It can be seen that the transmittable power rises with the voltage only up to definite limits depending on the type of dielectric but falls again beyond these limits[1]. It is obvious that the operation of a cable system at voltages above the power maximum would not make economic sense. The higher the dielectric loss coefficient $\varepsilon_r \cdot \tan\delta$ of the insulation and the smaller the thermal power dissipation P'_{ab}, the sooner the corresponding limit voltage is reached. Figure 3.28 shows this relationship as a comparison of the voltage limits of the four high-voltage dielectrics at different levels of heat dissipation. It can be seen that, on thermal grounds, cables for transmission voltages above the presently installed level of 500 kV are only possible without forced cooling by using XLPE and PPL.

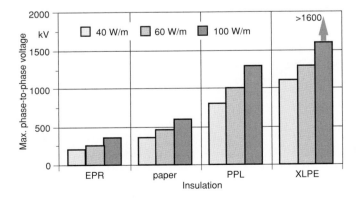

Figure 3.28
Transmission voltage limits

---

[1] It should be mentioned for completeness that the current is limited to a maximum of 1100 A/phase, the rated value for 1600 mm² Cu, even in areas of low dielectric losses.

3.5 Current Carrying Capacity of Naturally Cooled Cables

It was already shown in the previous section that the upper transmission voltage limit of cable systems is restricted principally by the dissipation of the heat losses occurring in the system. The same consideration applies to the maximum current carrying capacity (see Equations 3.27, 3.29). This is also the result of a thermal balance in which the temperature rise due to losses in the cable may not exceed an upper limit determined by the type of insulation material (see Table 3.2). What temperature actually occurs at which part of the cable is not just a question of the type of cable, system configuration and maximum load but depends to an almost greater degree on the environment in which the cable is operated and how the load varies with time.

3.5.1 Cable Temperature Rise

The cause of the temperature rise is the voltage-dependent and current-dependent losses in the various components. Since the voltage-dependent (dielectric) losses for a particular dielectric and constant transmission voltage are of a constant value, the total value of the losses can only be influenced by varying the load current. As the temperature rise process is subject to considerable time constants because of the thermal storage capacity of the cable and, if applicable, that of the surrounding soil, the maximum current carrying capacity also depends on the *load factor* of the system.

Load factor

The consumption of electrical power is subject to definite variations with time as shown by characteristic annual, weekly and daily load cycles. Figure 3.29 shows an example of a daily load curve that might occur in a mixed consumer structure including industrial and commercial premises and private households. The average load, the load factor m, of a transmission line over 24 hours corresponds to the area under the load curve and can be calculated as

$$m = \frac{1}{I_{max} \cdot 24\,\text{h}} \cdot \int_0^{24\,\text{h}} i(t)\,dt \tag{3.33}$$

where I_{max} is the maximum value of the current within a 24 h period.

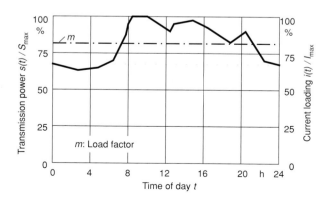

Figure 3.29
Daily load curve of electrical power consumption (principle)

Considerable differences in load during a day correspond to low values of m and lower loading of the transmission line. This can be taken into account when designing cable systems by using a smaller conductor cross-section for a particular maximum current than would be necessary with uniform loading [35]. In the past, this was the preferred method of sizing medium voltage cables in particular, generally taking a load factor of 0.7. However, in recent years the power supply companies have made increased efforts towards uniform loading of their systems. The average loading of the German power supply networks at the high-voltage level is now already of the order of 0.9 or even higher, so that new high-voltage cables are usually designed for full load ($m = 1$).

Internal and external thermal resistances

The heat generated within the various components of a cable is dissipated to the surroundings if there is no forced cooling. A *heat flow* N_v in W/m is thus created that flows from the respective *heat source* under the influence of a temperature gradient $\Delta \vartheta$ via a series of *thermal resistances* T to the *heat sink* in the atmosphere. According to whether these thermal resistances are formed from parts of the cable system itself or from areas in the surroundings, they are called internal or external thermal resistances. Their dimensions are of the form $K \cdot m/W$.

- *Internal* thermal resistances include all the non-metallic components of the cable, i.e. the insulation including the field smoothing layers, possible padding layers in the screen/sheath area, polymer sheath and corrosion protection and, in the case of three-core cables, the pressurized gas or pressurized oil filling in the common steel pipe. Metallic components are generally regarded as isothermal because of their heat conducting properties.

- *External* thermal resistance is formed mainly by the surrounding soil in the case of buried cables and by the interface between the cable surface and the air in the case of cables laid in air. In the latter case, heat dissipation takes place by *convection* and *radiation*.

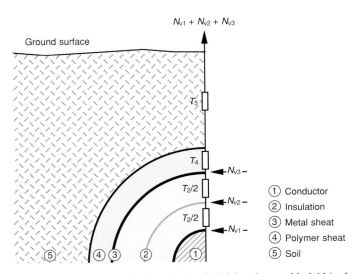

Figure 3.30 Thermal equivalent circuit of a high voltage cable laid in the ground

Analogous to electrical technology, the individual thermal resistances can be combined into a *thermal equivalent circuit* in that the heat flows are fed into the associated loss sources. The resulting temperature differences can then be calculated step by step using the "*Ohm's law of heat conduction*"

$$\Delta \vartheta = N_v \cdot T \tag{3.34}$$

As a simple example, Figure 3.30 shows the thermal equivalent circuit of a buried high-voltage cable with metallic sheath. The loss sources to be taken into account as injected heat flows are the conductor (N_{v1}), the metallic sheath (induced sheath currents N_{v3}) and the insulation (N_{v2}). The dielectric losses produced in the insulation are largely uniform in the whole of the dielectric. However, to make it possible to represent the system in a numerical calculation, the loss source is considered to be an endless thin cylinder concentrated at the physical center of the insulation layer; the heat flow N_{v2} originating from this therefore only has to overcome half of the associated thermal resistance T_2 [8].

From this equivalent circuit and in conjunction with Equation 3.34, the conductor temperature rise $\Delta \vartheta_L$ compared to atmospheric temperature can be determined as

$$\Delta \vartheta_L = (N_{v1} + 0.5\ N_{v2}) \cdot T_2 + (N_{v1} + N_{v2} + N_{v3}) \cdot (T_4 + T_5) \tag{3.35}$$

and in the same way, the temperature difference cable surface/atmosphere $\Delta \vartheta_o$:

$$\Delta \vartheta_o = (N_{v1} + N_{v2} + N_{v3}) \cdot (T_4 + T_5) \tag{3.36}$$

Nevertheless, a numerical calculation requires both the individual losses (Section 3.4) and the various thermal resistances to be known. As far as internal thermal resistances are concerned, the relationships between material and geometry are clearly specified. For the present example of a single core cable with its cylindrical components, the following applies, analogous to electrical resistance

$$T = (\rho / 2\pi) \cdot \ln (D / d) \tag{3.37}$$

where

D, d = external and internal diameters of the component in question
ρ = specific thermal resistance of the material used in $K \cdot m / W$

For most insulation and sheath materials, the specific thermal resistances lie relatively close together at 5 to 6.5 $K \cdot m / W$; only polyethylene and XLPE with values around 3.5 $K \cdot m / W$ deviate significantly from the values for the other materials (Table 3.8).

The relationships are more complex when determining the external thermal resistance of buried cables. This is because the specific thermal resistance of the soil as the most important parameter depends not only on the composition of the soil (which can vary considerably along the cable run) but also on the current condition of the soil which, in turn, is influenced by the operation of the cable. The most important factor here is the moisture content of the soil, which can change the specific thermal resistance by several hundred

Table 3.8
Specific thermal resistance ρ of some insulation, sheath and corrosion protection materials.

Material	impr. paper	PE, XLPE	EPR	PVC	impreg. jute
ρ in $K \cdot m / W$	5.0...6.5	3.5	5.0	6.0	6.0

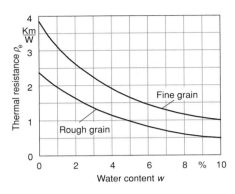

Figure 3.31
Specific thermal resistances of different moist
sandy soils (according to [8])

percent: the dryer the soil, the higher its thermal resistance as shown in Figure 3.31 with
an example of two sandy soils. The thermal design of the cable system must therefore take
into account the most unfavorable geological and meteorological conditions and, in addition, the possibility that the cable operation itself leads to further drying of the soil (more
information in the next subsection). *Without* additional drying, the specific thermal resistance of moist soil in Germany can generally be assumed to be 1 K·m/W.

Independent of the specific properties, the calculation of heat transport through the soil
also requires its geometric thermal resistance T_{e0} to be known. For the example of the
single core cable from Figure 3.30, a mirroring method is used (Figure 3.32). The cable
K with external diameter D_K lying at a depth h is regarded as an endless cylinder and is
mirrored at the ground surface so that a further endless cylinder K' exists at a distance h
above the surface of the ground. This forms the heat sink into which the heat from the
cable K flows. Using an analogy to the laws of electrostatics and with the approximation
$h \gg D_K$ results in the following geometric thermal resistance of the soil in this configuration (other laying configurations can also be approximated using electrostatic calculation methods):

$$T_{e0} = (1/2\pi) \cdot \ln (4h/D_K) \tag{3.38}$$

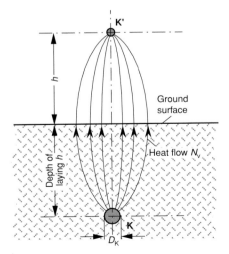

Figure 3.32
Derivation of the thermal
resistance of the soil

and then the actual external thermal resistance T_5 from Figure 3.30:

$$T_5 = \rho_e \cdot T_{e0} = (\rho_e / 2\pi) \cdot \ln(4h / D_K) \tag{3.39}$$

where ρ_e is the specific thermal resistance of the soil in accordance with Figure 3.31.

Equation 3.39 also shows that the external thermal resistance reduces with increasing external diameter of the cable, a relationship that is sometimes used in modified form to specifically improve the natural cooling (see Section 3.5.2).

If the cable is laid in air, the thermal resistance of the soil is replaced by the *heat transfer resistance* T_T from the cable surface to the surroundings. It is calculated per unit length as

$$T_T = 1 / (\alpha \cdot \pi \cdot D_K) \tag{3.40}$$

where α is the heat transfer coefficient in $W/(m^2 \cdot K)$.

The *heat transfer coefficient* depends partly on the sheath material, its color and surface properties. As a rule, the resulting heat transfer resistance is considerably smaller than the thermal resistance of the soil, so that cable laid in air can be thermally loaded to a significantly higher degree than the same configuration in the ground.

Drying-out of the soil

It can be seen from Figure 3.31 that the moisture content of the soil has a considerable influence on its thermal conductivity. It was also noted that this condition could be strongly influenced by the operation of the cable itself. In fact, a thermal gradient $\vartheta(x)$ develops starting from the conductor temperature via the insulation and the sheath of the cable that continues in the adjoining soil and (theoretically) only reaches the temperature of the undisturbed soil at an indefinite distance (see Figure 3.33). The steepness and shape of this temperature profile is determined by the values of the individual specific thermal resistances, analogous to the voltage drop along a series of electrical resistances.

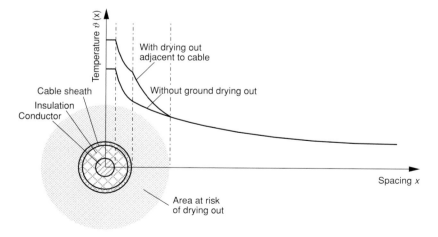

Figure 3.33 Temperature profile around a single core cable (schematic)

Because of the increased temperature at the surface of the cable, the partial pressure of the water vapor rises in the immediate surroundings. The pressure drop that arises creates a flow of moisture away from the cable to outer areas where the water molecules condense again into droplets when the temperature falls below the *dew point*. The result of this is the continued drying-out of the soil adjacent to the cable, combined with a corresponding increase in the specific thermal resistance of the soil in this area. At the same time, the temperature profile changes; the curve flattens at greater distances (Figure 3.33).

The process described here usually takes several weeks or even months until a stable final state is reached. Three characteristic conditions are possible, depending on the soil composition, the frequency of precipitation and the amount of cable losses:

- The moisture content and therefore the thermal resistance in the vicinity of the cable remain largely unchanged; significant drying of the soil does not occur. This condition is to be expected in zones of high precipitation with soils containing lime or with low cable loadings. It represents a thermally stable total system.

- The vaporization is so intense that the soil dries out even at large distances from the cable and therefore its thermal resistance is significantly increased. The heat dissipation, significantly reduced by this effect, causes additional temperature rise in the cable and further expansion of the dry zone. The system is thermally unstable; accelerated ageing of the cable insulation right through to thermal breakdown can be expected. This scenario is enhanced by low precipitation, poor soil and high cable loading. For example, sandy soils tend to dry out with cable surface temperatures as low as 30 °C [8].

- The vaporization is sufficient to cause drying of the soil and limited temperature rise in the immediate vicinity of the cable. However, the expansion of the dry zone reaches only a few cm, so that the system in total remains thermally stable This condition, known as *controlled soil drying*, is sometimes induced deliberately so that, especially in the case of XLPE cables, the maximum permissible conductor operating temperature of 90 °C can be fully utilized and thus the heat dissipation according to Equation 3.35 can be specifically increased [36]. The dry area is then described by an isothermal that is eccentric to the cable, the *boundary isotherm* for e.g. 15 K temperature difference to the cable surface. Within this boundary isotherm, the specific thermal resistance of nearly completely dry sandy soil ($\rho_e = 2,5$ Km/W) is assumed.

Taking accessories into account

Whereas the temperature inside the insulation of radial field cables reduces logarithmically from inside to outside in a similar way to the electrical field strength, this distribution is, in most cases, noticeably distorted in the area of the accessories. As described in detail in Section 6, this is due to the fact that, with few exceptions, considerably greater thicknesses of insulation are required in the accessories than in the cables themselves to achieve the same dielectric strength [26, 37]. Assuming the same dielectric loss coefficient as in the cable insulation, and using Equations 3.30 and 3.31 from Section 3.4.2

$$P'_{vu} = U^2 \, \omega \, C'_0 \cdot \varepsilon_r \cdot \tan\delta \tag{3.30}$$

with

$$C'_0 = (2 \, \pi \, \varepsilon_0) \, / \, \ln(D \, / \, d) \tag{3.31}$$

and although this actually leads to a reduction of the voltage-dependent (dielectric) losses because of the lower vacuum capacitance C_0', at the same time the internal thermal resistances of the insulation increase and these have to be overcome by the current-dependent conductor losses, which are in any case dominant at full load. This means that, especially in the case of joints, a locally increased conductor temperature must be included in the calculation unless special measures have been taken to extract heat from this locality. The longer the joint and the smaller the conductor cross-section, the more pronounced this effect becomes; on the other hand, in short joints with large conductor cross-sections, a degree of axial temperature equalization occurs because of the good thermal conduction properties of the conductor.

Limited axial temperature increases can also occur at other points along a cable run, and measures must be taken to counter these in the interest of higher power transmission capacity: at crossing points with other equally highly loaded cable systems or district heating pipes or in areas where account must be taken of unsatisfactory re-moisturizing of the soil because of building conditions, leading to greater external thermal resistance [38, 39]. These include e.g. wide, fully sealed asphalt roads.

3.5.2 Ways of Increasing the Power Transmission Capacity

The most important parameters that influence the power transmission capacity of a cable run are naturally the rated voltage and the conductor cross-section. But, even after this base data has been specified, there is still a whole range of possibilities to increase the power transmission capacity at the planning stage. This begins with the choice of conductor design and dielectric which allows the reduction of the current-dependent and voltage-dependent losses (see Sections 3.4.1 and 3.4.2) and, at the same time, the specific internal thermal resistances (Table 3.8). Further possibilities to reduce the operating losses exist in the use of one-end-grounded or cross-connected cable sheaths to avoid conductor loops in the area of the metallic coverings (Figure 3.22) and in the maintenance of a minimum spacing between the individual phases to exclude the increase of the conductor resistance due to the proximity effect [6].

All other measures to increase the power transmission capacity of naturally cooled cable systems are aimed at reducing the external thermal resistances and thus improving the dissipation of heat from the cable. In detail, these are:

- Thermally stable trench backfilling to prevent drying-out of the soil over the whole cable run;

- Thermally stable trench backfilling at selected *hot spots*, i.e. in the vicinity of joints or crossing points with other cables or district heating pipes;

- The use of *heat pipes* (heat extraction conductors) in the area of specific hot spots.

"Thermally stable backfilling" means the partial or complete replacement of the excavated soil by bedding material that maintains a comparatively low thermal resistance ρ_E under full load and at high temperature (load factor $m = 1$). The aim is to achieve values of $\rho_E \leq 1$ Km/W even after full drying-out and this is best achieved by the use of particular sand-gravel mixtures with limestone flour or *lean concrete* with a low cement content ([8], S. 227ff.).

Provided the choice of filling material and the dimensions of the thermally stable back-

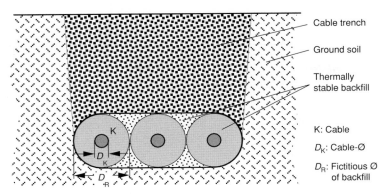

Figure 3.34 Cable system laid in thermally stable bedding material

fill area are correct , this step not only prevents the controlled or even uncontrolled dry-ing-out of the soil as described previously, but the thermal resistance of the remaining soil is also reduced – and this independent of its specific heat conduction properties. If a single-core cable with diameter D_K as in Figure 3.32 is again assumed and if, as an additional approximation, the thermally stable backfill area is assumed to be a concentric circle with diameter $D_R > D_K$ around the cable (Figure 3.34), the reduction from immediately above of the thermal resistance of the soil remaining outside the backfilling can be obtained from Equation 3.39:

$$T' = (\rho_e / 2\pi) \cdot \ln (4h / D_R) < T = (\rho_e / 2\pi) \cdot \ln (4h / D_K) \tag{3.39}$$

where T, T' are the thermal resistances without and with thermally stable backfilling.

While the improvement of heat dissipation described here can, if required, be used over the whole cable run or be limited to specially thermally-stressed sections, heat pipes are basically only suitable for the thermal relief of critical hot spots. These elements are fluid-filled copper or aluminium pipes that lie in as direct as possible contact parallel to the cable or accessory in the area at risk over a few meters of length and discharge the heat absorbed there to a distant area of the soil (see Figure 3.35).

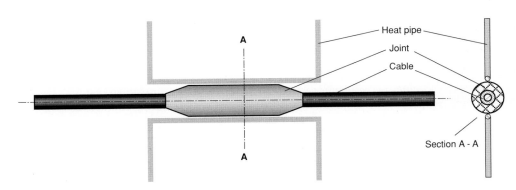

Figure 3.35 Arrangement of heat pipes for extracting heat from hot spots

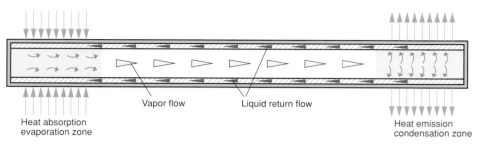

Vapor flow   Liquid return flow

Heat absorption
evaporation zone

Heat emission
condensation zone

Figure 3.36   Functional principle of heat pipes

The basis of the high heat transport capacity of heat pipes is there liquid filling with a specially determined boiling point, similar to that used in heat pumps or refrigeration units. If chosen correctly, the liquid vaporizes at the heat source (e.g. the joint), flows to the area of low temperature, condenses here, giving up its heat of condensation to the surroundings and then flows back again to the heat source [40]. Figure 3.36 shows this closed cooling circuit that operates without moving parts or external power and therefore forms a completely maintenance-free system that will remain functional over the whole life of the cable. The internal temperature differences remain within such narrow limits that the heat pipes represent an isothermal over practically their whole length at a level between the cable surface temperature *without* heat extraction and that of the distant soil area where the losses are to be dissipated [41].

A similar effect can be achieved without heat pipes at local hot spots in low-viscosity impregnated paper cables with hollow conductor. This is done by deliberately allowing the impregnating medium to flow to and fro (oscillate) between the expansion tanks at both ends of the cable system and thus extract the additional heat loss from e.g. the area of a joint to the adjacent cable section that is not thermally overloaded. Nevertheless, this requires the use of an oil pump and thus represents a preliminary step towards a high power cable with forced cooling which will be fully described later (Section 5.1).

3.5.3 Load Capacity Calculations

To determine the load capacity of naturally cooled cable systems, a whole range of limiting conditions must be observed and specified, especially the laying conditions and the soil properties. The data of the cable itself also naturally play a decisive role – type of dielectric, rated voltage, conductor cross-section – as shown in the following calculation examples.

Influence of rated voltage and conductor cross-section on the load capacity

The load capacities per system for single core cables laid direct in the ground with thermally stable backfilling, controlled drying of the soil outside the backfilling, cables laid side by side with on 300 mm centers, cross-bonding of the metallic sheath and load factor 1 for voltages U_N = 110 kV, 220 kV and 400 kV are calculated according to the conductor cross-section and shown together in Figure 3.37. It can be seen from this that, with natural cooling and based on the above limiting conditions, the maximal system load capacity that can be achieved is about 200–250 MVA at 110 kV, 450–550 MVA at 220 kV and 600–850 MVA at 400 kV.

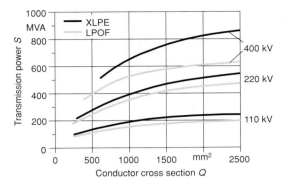

Figure 3.37
Transmission capacity per system of oil-impregnated paper-insulated and XLPE cables in relation to rated voltage and conductor cross-section (see text for limiting conditions)

Thermal capacity limits

It was shown in the previous sections that both the highest operating voltage and the maximum current carrying capacity of naturally cooled cable systems and thus their load capacity is limited mainly by the amount of heat losses that can be dissipated. In the normal climatic conditions in Germany, the application of all the possibilities described above for the reduction of the effective thermal resistances and, at the same time, the prevention of local hot spots results in a best case heat dissipation rate of 90 W/m [42]. If this value is used at the thermal power P'_{ab} and taking the dielectric data ε_r and $\tan\delta$, from Table 3.7, the thermal capacity limit S_{therm} for naturally cooled cable runs with various types of insulation can be approximately calculated using Equation 3.32:

$$S_{therm} = \frac{U}{\sqrt{R'}} \cdot \sqrt{P'_{ab} - U^2 \omega C'_0 \cdot \varepsilon_r \cdot \tan\delta} \tag{3.32}$$

Figure 3.38 summarizes the results at $U_N = 400$ kV and 500 kV in the form of a bar chart based on a 2500 mm² copper Milliken conductor with individual wire insulation as a low-loss design variant for all three types of cable (oil-impregnated paper, XLPE and PPL) evaluated here.

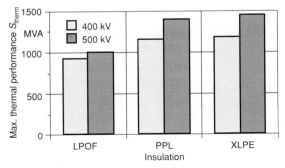

Figure 3.38
Thermal power limits of naturally cooled extra high voltage cable systems with 2500 mm² Cu conductor with a heat dissipation of 90 W/m

This shows that maximum load capacities per system of between approximately 900 MVA (400 kV oil-impregnated paper-insulated cable) and 1450 MVA (500 kV XLPE cable) are possible without forced cooling (see Table 1.1). The values are considerably higher than the load capacities of naturally-cooled cables with the same cross-section but without optimization of heat dissipation (e.g. Figure 3.37), but still only reach about 50 % of the load capacity of overhead lines operating at the same voltage [3, 42]. This means that, if the full capacity of a 400 kV or 500 kV overhead line has to be catered for in a mixed transmission route, at least a double system or the use of forced cooling is needed, or perhaps a completely different cable technology (see Section 5) [43].

3.6 Transmission Characteristics of Cables

3.6.1 Iterative Network, Cable Parameters Per Unit Length

The operating properties of an extended cable route can no longer be correctly predetermined just on the basis of concentrated circuit elements. Rather, the cable should be regarded as a series connection of many *line elements* of length dx that consist of respectively identical circuit elements $R'dx$, $L'dx$, $C'dx$ and $G'dx$ as shown in Figure 3.39. The determining values R', L', C' und G' of the individual cable elements are known as the *cable parameters per unit length* and can be derived from the geometry of the cable together with the electrical properties of the materials used. The series connection of n such line elements form an *iterative network* whose operating behavior can be calculated by solving the *general telegraph equation*

$$\frac{\delta^2 u}{\delta x^2} = R'G' \cdot u + (R'C' + L'G') \cdot \frac{\delta u}{\delta t} + L'C' \cdot \frac{\delta^2 u}{\delta t^2} \tag{3.41}$$

taking the respective limiting conditions into account. In addition to the number of line elements and the values of the cable parameters per unit length as a measure of the geometric length, the determining factors here are the form and frequency of the applied voltage and the *electrical length* of the iterative network, derived from all the parameters together. The order of size of the cable parameters per unit length of high-voltage and extra high-voltage cables and the impedances at 50 Hz derived from these are shown in Table 3.9 and their qualitative dependence on the conductor cross-section at constant operating voltage is shown in Figure 3.40. Detailed information on the calculation of the parameters per unit length for different cable system configurations can be found in [8].

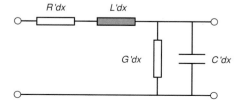

Figure 3.39
Equivalent circuit of a cable element of length dx

Table 3.9 Cable parameters and impedances per unit length of high voltage and extra high voltage cables at 50 Hz

R' in Ω/km	L' in H/km	$\omega L'$ in Ω/km	G' in $1/(\Omega$km$)$	C' in F/km	$\omega C'$ in $1/(\Omega$km$)$
$0.01\ldots0.05$	$0.3\ldots0.7\cdot10^{-3}$	$0.1\ldots0.2$	$0.1\ldots0.5\cdot10^{-6}$	$150\ldots300\cdot10^{-9}$	$50\ldots100\cdot10^{-6}$

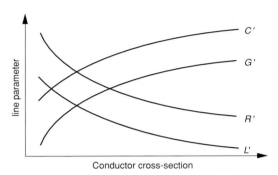

Figure 3.40 Influence of conductor cross-section on the cable parameters (schematic)

3.6.2 Line Equations, Cable Parameters and "Electrical Length"

Provided that the voltages and currents in the line are sinusoidal, the solution of the telegraph equation 3.41 leads to the so-called *line equations*. Written in complex form, these are:

$$\vec{U}(x) = \vec{U}_2\cosh(\gamma x) + \vec{I}_2\vec{Z}_w\sinh(\gamma x) \tag{3.42}$$

and $\quad \vec{I}(x) = \vec{I}_2\cosh(\gamma x) + (\vec{U}_2/\vec{Z}_w)\cdot\sinh(\gamma x) \tag{3.43}$

In the above: $\vec{U}_2, \vec{I}_2$ = the complex currents and voltages at the end of the cable,
$\qquad\qquad\quad \vec{Z}_w$ = the (also complex) *characteristic impedance* of the cable,
$\qquad\qquad\quad \gamma$ = the so-called *propagation constant* and
$\qquad\qquad\quad x$ = the distance to the end of the cable.

$\vec{Z}_w$ and γ are part of the *cable parameters,* also known as "secondary cable parameters" that can be calculated directly from the parameters per unit length:

$$\vec{Z}_w = \sqrt{\frac{R' + j\omega L'}{G' + j\omega C'}} \tag{3.44}$$

and $\quad \gamma + \sqrt{(R' + j\omega L')(G' + j\omega C')} = \alpha + j\beta \tag{3.45}$

where α is the *attenuation constant* and β is the *phase constant.*

The attenuation and phase constants are also part of the cable parameters group that in turn can be derived from the parameters per unit length:

$$\alpha = \sqrt{\frac{1}{2}(R'G' - \omega^2 L'C') + \frac{1}{2}\sqrt{(R'^2 + \omega^2 L'^2)(G'^2 + \omega^2 C'^2)}} \tag{3.46}$$

und $\quad \beta = \sqrt{\frac{1}{2}(\omega^2 L'C' - R'G') + \frac{1}{2}\sqrt{(R'^2 + \omega^2 L'^2)(G'^2 + \omega^2 C'^2)}} \tag{3.47}$

The phase constant affects two other characteristic values that influence the operating behavior of a power cable represented as an iterative network: the *phase velocity v* and the *wavelength λ* of an electromagnetic wave in the cable under consideration

$$v = \omega / \beta \tag{3.48}$$

and $\qquad \lambda = v \cdot T = 2\pi / \beta$ (3.49)

where $T = 2\pi/\omega$ is the period of oscillation.

The relationship of wavelength λ and geometric length *l* of the cable finally determines whether the cable being represented as an iterative network is electrically "short" or "long" and how the individual cable elements are best arranged for the estimation of the operating behavior of the cable. In this context, the following apply

- $l \le \lambda/30$: *electrically short cable*; this can usually be represented sufficiently accurately by concentrated elements

- $l > \lambda/30$: *electrically long cable*; for this, the complete iterative network equivalent circuit with *n* distributed elements must be used.

3.6.3 Equivalent Circuit and Operating Parameters

All the cable parameters introduced in the previous chapter only become sufficiently easy to grasp if a few approximations and simplifications are allowed. Table 3.9 indicates that the ohmic transverse conductance G' with values of $0.1...0.5 \times 10^{-6}$ $(\Omega \text{ km})^{-1}$ can be neglected in comparison with the capacitive leakance $\omega C'$ with values of $50...100 \times 10^{-6}$ $(\Omega \text{ km})^{-1}$ in the formulae for determining the various parameters of high-voltage cables. In a first approximation, this also applies to the longitudinal impedances R' and $\omega L'$, only the inductive component of these usually being taken into account when estimating parameters according to Equations 3.44 to 3.49. This means that, unlike the case of the thermal balance discussed earlier, a cable route is regarded for this purpose as a *loss-free line*.

With these approximations, the parameter equations from Section 3.6.2 are simplified to produce the formulae summarized in Table 3.10 that finally allow the numerical estimation of the determining key values of the operating properties of the cable route at a given frequency. Table 3.11 shows the guide values for these at 50 Hz.

The wavelength λ of about 3000 km characterizes all cables with a length $l \le 100$ km as "electrically short" cables on the basis of the limit λ/30 mentioned in the previous section, so that practically all three-phase cable systems can be assigned to this category (see Section 2.2.1, Figure 2.1). With this prerequisite, the iterative network with *n* cable elements in Figure 3.39 can be simplified as a π equivalent circuit with concentrated elements as shown in Figure 3.41. The values of the individual circuit elements are calculated from the respective parameters per unit length of the iterative network (Table 3.9)

Table 3.10 Line parameters for loss-free lines

Attenuation-constant	Phase-constant	Propagation-constant	Characteristic impedance	Phase-velocity	Wavelength at frequency f
$\alpha = 0$	$\beta = \omega \sqrt{L'C'}$	$\gamma = j\omega \sqrt{L'C'}$	$Z_W = \sqrt{L'/C'}$	$\nu = 1/\sqrt{L'C'}$	$\lambda = 1/(f \cdot \sqrt{L'C'})$

Table 3.11
Guide values of some key operating values of high voltage and
extra high voltage cables at 50 Hz

Characteristic impedance Z_w	Phase velocity ν	Wavelength λ
50 Ω	150.000 km/s	3000 km

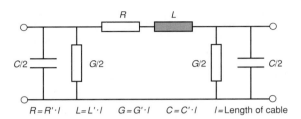

$$R=R'\cdot l \quad L=L'\cdot l \quad G=G'\cdot l \quad C=C'\cdot l \quad l=\text{Length of cable}$$

Figure 3.41
Representation of a cable as an electrically short line using a π equivalent circuit with
concentrated elements

multiplied by the total length l of the cable, i.e. $R = R' \cdot l$, $L = L' \cdot l$, $C = C' \cdot l$ and
$G = G' \cdot l$. Transverse capacitance C and transverse conductivity G are now each concentrated half at the beginning and half at the end of the cable and thus appear in the
π circuit as $C/2$ and $G/2$ respectively before and after the series connected longitudinal
impedances R and L.

As the voltage-dependent losses that occur in the insulation layer cannot be neglected,
even with high grade dielectrics such as XLPE and PPLP and especially at extra high-
voltages (see Table 3.7), the transverse conductivity value G representing the dielectric
loss cannot be left out of the cable equivalent circuit, unlike the case when calculating
the cable parameters.

The equivalent circuit developed in this way provides a comparatively simple way to
estimate the operating behavior of cable routes. For instance, taking into account the
guide values for the individual circuit elements given in Table 3.9, it can be seen immediately that cables represent a predominantly capacitive load in the network that must be
compensated inductively at a particular length. In addition, it is clear that the resistive
discharge current flowing via the conductance G, as a measure of the dielectric losses,
depends only on the operating voltage, while the conductor losses converted to heat in
the longitudinal resistance R depend only on the load current. Finally, using the equivalent circuit, the currents and voltages, their phase angle and changes under various load
conditions can also be calculated exactly or shown visually as a vector diagram [8].

3.7 Cables for High-Voltage DC Transmission (HVDC)

All the statements made previously about the operating behavior of high-voltage and
extra high-voltage cables refer to their application in single-phase and three-phase AC
networks with frequencies of 50 or 60 Hz. Compared to these, characteristic differences

occur on physical grounds when high DC voltages are applied and these must be taken into account for HVDC transmission applications. The most important characteristics of cables under DC voltage stress are:

- There is no capacitive charge current; the length limitation as shown in Figure 2.1 does not apply.

- Instead of the dielectric losses, only the much lower resistive leakage through the insulation needs to be taken into account.

- Current displacement and sheath voltage induction do not exist as there are no changes of the magnetic field with time. As a result, there are no sheath losses and the conductor cross-section is utilized to its full extent.

- The field distribution within the dielectric is subject to a purely resistive control by the specific DC conductivity of the insulation rather than the capacitive control in the case of AC operation.

- Partial discharges in defects in the dielectric occur – if at all – only extremely rarely as the periodic recharging effect due to the sinusoidal stress voltage is not present.

3.7.1 Areas of Application for High-Voltage DC Transmission

In spite of these obvious advantages of DC transmission by cable, this technology has to date only made its mark in special applications. These include. in the first place, very long transmission routes such as are necessary for e.g. submarine cable links which cannot be operated with AC because of the limitations of capacitive charging current already discussed. As a result, most of the HVDC cable routes implemented up to now are in Scandinavia [44]. Forward-looking projects are at present being proposed for submarine cable links between Italy and Greece [45] and between the Japanese islands of Honshu and Shikoku [46].

The requirement for HVDC submarine cables up to the year 2005 is estimated to be a total of about 5000 km, about half of this in Europe and half in the ASEAN states. Noteworthy projects among these are the Bakun project in Malaysia with a requirement for around 3 x 600 km of cable, three projects each about 500 km long between Norway and Germany and the Netherlands and a planned link between Iceland and Scotland with a length of about 1000 km. HVDC cables are therefore the growth area in power cable technology.

Conversely, overland links using HVDC cables are a rarity worldwide at present. The most important route of this type is the 640 MW link between Kingsnorth and London which has been in operation since 1975 [47]. The reason for the low growth in the use of HVDC overland cables in spite of their low losses is mainly the economic facts.

- In the highly interconnected city power supply networks, the preferred area of application for cables, HVDC cannot be considered as, up to now, only *A to B connections* without branches have been possible and the rectifier and inverter stations that are required would be unacceptably expensive and

- Overland long-distance links passing through areas of less dense population are, as in the past, preferably carried out as overhead lines because of the basic cost relationships to cable as already discussed in Section 1.3.1. These are equally valid whether the route is to be operated as a DC or an AC system.

Table 3.12 Key data of selected HVDC cable projects

Project	Voltage	Power	Route length	Cable type
Kontek	±400 kV	600 MW	170 km	Oil-impregnated paper insulated
Baltic Cable	±450 kV	600 MW	260 km	Mass-impregnated paper insulated
Bakun*	±500 kV	700 MW	670 km	Mass-impregnated paper insulated

*) being planned

Table 3.12 summarizes the most important data of three HVDC cable projects that have been implemented or are being planned.

3.7.2 Electrical field with DC Voltage Stress

Field control mechanisms

It has already been mentioned that the field distribution inside the dielectric when stressed by a DC voltage is subject to resistive rather than capacitive control. Thus, with a given voltage level and cable geometry, the specific DC conductivity κ rather than the permittivity ε_r is responsible for the value of the electrical field strength inside the insulation. Because of the analogy between the electrostatic and electrical flow field, the field strength is divided into a series circuit of insulation areas with differing permittivity values and conductivities in theoretically identical forms over the two areas 1 and 2 as shown in Figure 3.42:

$$E_1 / E_2 = \varepsilon_{r2} / \varepsilon_{r1} = \kappa_2 / \kappa_1 \tag{3.50}$$

This means that, in the case of DC voltage, there is a lower field strength in areas of higher conductivity in the same way that, in the case of AC voltage, there is a lower field strength in zones of high dielectric constant than in neighbouring areas with lower κ and ε_r.

This resistive control has two important consequences for cable technology:

- In graded insulation such as impregnated paper or PPLP, the higher stress from the impregnation medium zone is transferred into the layers of solid material, as the

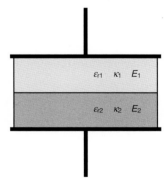

Figure 3.42 Different dielectrics in series connection

85

impregnation medium basically has a higher specific DC conductivity than cellulose or polypropylene and

- The field distribution is dependent on temperature since the DC conductivity of all the dielectrics in question rise exponentially with temperature according to the *Van't Hoff law*:

$$\kappa = \kappa_o \cdot \exp(-W/kT) \tag{3.51}$$

or
$$\ln(\kappa) = \ln(\kappa_0) - W/kT \tag{3.52}$$

where:
W = thermal activation energy
k = Boltzmann constant
T = abs. temperature in K
κ_0 = theoretical conductivity at 0 K.

The logarithmic form of the Van't Hoff law (Equation 3.52) means that there is a linear relationship between the logarithm of the specific DC conductivity and the reciprocal of the absolute temperature. Figure 3.43 shows a graph of this type for cable paper impregnated with high-viscosity polyisobutylene PIB, which is used extensively for the impregnation of mass-impregnated and external gas pressure cables (Section 4.1.5). According to this graph, the specific DC conductivity increases by an order of magnitude for each 25 K temperature rise.

On its own, this does not change the field distribution so long as the temperature rises and falls evenly in the whole dielectric. In the discussion on cable temperature rise in Section 3.5.1 it was, however, shown that the greatest source of losses in current-carrying cables is in the conductor and, as a result of this, a negative temperature gradient forms in the insulation layer from the inside to the outside (see e.g. Figure 3.33). Provided the conductor temperature is unchanged, and because of the lower dielectric losses, this effect is even more pronounced in the case of DC cables than under AC conditions. Temperature differences of ≥ 25 K between the insulation zones close to the conductor and those distant from the conductor are therefore entirely possible in thick-wall HVDC cables under operating conditions.

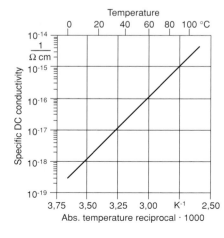

Figure 3.43
Influence of temperature on the specific DC conductivity of PIB-impregnated paper dielectric[48]

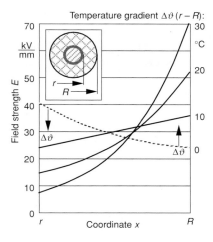

Figure 3.44
Field distribution in impregnated paper dielectric of 400 kV HVDC cables at various temperature differences between conductor and metallic sheath

This means that, during the process of warming up to rated conductor temperature that starts to take place following the commissioning of an HVDC cable, a difference in conductivity of more than one order of magnitude is created between the inner and outer areas of insulation. As a result of this, the electrical field strength near to the conductor reduces continuously in accordance with Equation 3.50 while, at the same time, the insulation near the sheath is more highly stressed. Figure 3.44 shows various states of the field distribution curve in the mass-impregnated paper insulation for a 400 kV HVDC cable [48]. Even with small temperature differences of about 10 K over the insulation layer, the location of the highest field strength moves from its original position next to the conductor into the boundary zones of the dielectric; further increases in conductor temperature reinforce this process considerably. With a temperature gradient of 20 K, a considerably higher electrical stress occurs in this example at the outer field limiting layer than occurred at the inner conductor with a cold cable. To combat any resultant over-stressing, the conductor temperature of HVDC cables is generally limited to substantially lower values than the materials employed would actually withstand. The permissible conductor temperature when using conventional paper dielectrics in HVDC cables is thus only around 50 °C instead of the 70 to 85 °C at which AC cables of the same design can be operated.

Special properties of extruded synthetic dielectrics

The stress distribution in solid polymer dielectrics at high DC voltage is even more complicated. In addition to the temperature influence that occurs in a similar manner to that described above, there are two further effects: change in field due to space charging and due the local field strength itself.

The charge carriers (electrons) that are injected from the negative electrode in the strong DC field tend to be stored in the insulation volume, especially in the case of highly insulating, low-loss materials such as polyethylene and XLPE [49, 50]. The carriers form the source of an additional space charge field that is superimposed on the geometrically determined field pattern. This intrinsic field is of low importance as long as the polarity of the applied voltage remains unchanged. If, however, rapid polarity reversals occur due to external overvoltages or switching events, the space charge field leads to more or less incalculable over-stressing whose effects on the long-term dielectric strength of the polymer have not yet been able to be clarified. In this context, reference should be made to the discussion that has lasted for many years about the danger of DC tests for PE/XLPE medium voltage and high-voltage cables [51].

Unlike the space charges, the locally-occurring field strength does not have an immediate effect on the field distribution but – in a similar way to the temperature – acts via the specific conductivity that also depends partly on the electrical stress. Without addressing here the complex physical mechanism that is responsible for this [13, 15, 50, 52–54], the theoretical relationship between conductivity κ and DC field strength E will simply be stated:

$$\kappa \sim E^m \text{ with } m > 0 \tag{3.53}$$

or $\quad \log(\kappa) = m \cdot \log(E) \tag{3.54}$

Figure 3.45 shows graphs of values measured on thin-wall test samples of thermoplastic low-density polyethylene (LDPE). It can be seen that, according to temperature, the conductivity begins to rise with field strength at 1 to 5 kV/mm and then rises by several orders of magnitude between about 10 and 50 kV/mm. This steep rise of conductivity that in turn affects the field strength in accordance with Equation 3.50 lies exactly in the range of operating stress that is usual in present-day HVDC cables. Although a theoretically similar effect applies also to oil-impregnated paper dielectrics, its effect there is less pronounced because of the considerably higher basic conductivity of the insulation [55].

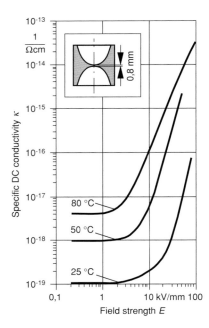

Figure 3.45
Influence of field strength on the DC conductivity of LDPE

This means that, unlike multi-layer dielectrics in which the field pattern only depends on the temperature distribution within the insulation layer, the stress distribution in extruded solid dielectrics such as PE and XLPE depends on a very complex mutual influence of temperature, space charge development and and local field strength. Since there is additionally a pronounced long-term effect, especially in the case of the space charge processes, the forecasting of the stress relationships that actually occur in polymer dielectrics under given external conditions is a very uncertain science.

As a result of this knowledge, current research on improving XLPE for HVDC applications is concentrated on giving the solid polymer dielectric a higher conductivity by using polarized or weakly conductive additives. This development work promises success, but has not yet progressed beyond a short laboratory prototype cable for 250 kV DC [56]. HVDC cables with solid polymer dielectrics have yet to be used in practice.

4 Types of High-Voltage and Extra High-Voltage Cables

Section 2.2.3 focused on the principal features of the two groups of dielectrics which can be used for the insulation of high-voltage and extra high-voltage cables: paper tapes saturated in suitable impregnating media of varying viscosity and lapped in multiple layers, and polymer materials based on polyethylene or synthetic rubber (EPR), which are applied to the conductor in a single layer by means of extrusion.

4.1 Cables with Impregnated Paper Insulation

4.1.1 Design of the Multi-Layer Dielectrics and Properties of the Individual Components

Multi-layer dielectrics normally consist of two components: paper and impregnating medium. In the case of low-loss PPL insulation (Section 3.4.2), a further component is added in the form of polypropylene, and in internal-gas pressure cables with a non-draining compound, gas-filled cavities may constitute a third component of the insulation (Section 3.1.1). The electric and dielectric properties of the dielectric result from the interaction of the individual components and are determined in part by the volume percentages of these. To achieve satisfactory properties, paper and impregnating liquids must be subjected to certain physical-chemical processes during cable manufacture.

Papers

Cable insulating papers are a natural product and are obtained from high-grade spruce wood or fir wood ("high-grade" assumes slow tree growth and low-resin wood; both are to be found predominantly in northern latitudes). The main constituent is so-called sodium cellulose, a linear macromolecule, in which hydrocarbon rings are bound to one another by oxygen bridges, as shown in Figure 4.1.

The properties of paper which are relevant for its use as a cable dielectric are determined specifically by two features of the cellulose:

- the three hydroxyl (OH)-groups per hydrocarbon ring of the macromolecule and

- the fact that paper cellulose does not form a homogeneous solid, but is always in the form of a porous material with a greater or lesser air content.

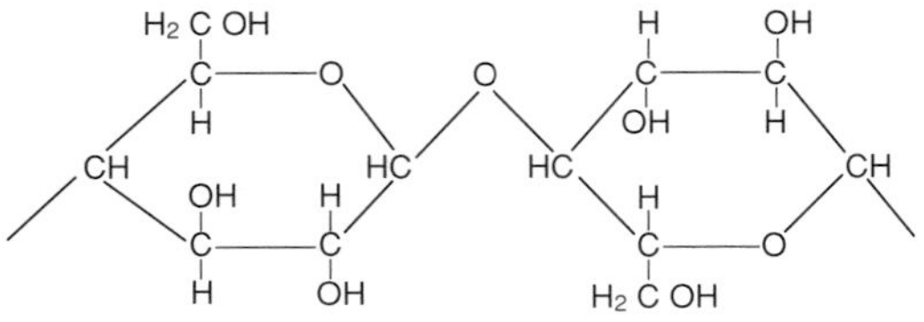

Figure 4.1 Cellulose molecule

90

The pores of the paper form so-called capillaries that tend to absorb fairly large amounts of moisture from the atmosphere (capillary effect = "blotting paper effect"), and the OH groups ensure that the absorbed moisture remains bound relatively firmly to the cellulose structure. Thus, normal cable paper absorbs up to 8% of its own weight when stored under normal indoor conditions (20 °C / 50% relative air humidity).

Since moisture has a particularly strong detrimental effect on the dielectric properties (loss factor, permittivity, DC conductivity) of the paper, as already illustrated in Section 3.2.2 (Figure 3.15), and also indirectly impairs the dielectric strength of the liquids used for impregnation, it must be carefully removed prior to impregnation, necessitating a lengthy and expensive vacuum process (see below) owing to the strong linkage to the OH groups.

Apart from the moisture content and purity, the so-called bulk density or weight by volume also influences the electrical properties of paper as a cable insulation to a considerable degree. This depends in turn on the cellulose content: a high cellulose content yields a high breakdown strength and high insulation resistance. On the other hand, the low porosity of highly dense paper gives rise to a high permittivity – high charging power and power loss – of the dielectric as a whole (permittivity of pure cellulose = 6.08) and a poor evacuation and impregnation capability, since under this condition only small flow cross-sections are available. The manufacturing process takes correspondingly longer.

Taking these contrary effects of increasing density on the electrical properties into account, papers with bulk densities of between approx. 0.8 and 1.2 g/cm³ , corresponding to a cellulose content of over 50% to just under 80% according to Figure 3.25, are used for high-voltage and extra high-voltage cables. To optimize the electrical properties, insulation in the extra high-voltage range ≥220 kV is built up of papers of varying bulk densities, with highly dense and yet extremely thin papers being used close to the conductor, whereas papers with a lower proportion of cellulose and a greater layer thickness are used in the remaining dielectric [26]. It is thus possible to reduce the field strength

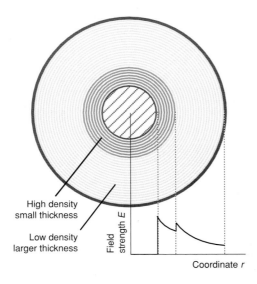

High density
small thickness

Low density
larger thickness

Field strength E

Coordinate r

Figure 4.2
Field-strength distribution in cable dielectrics with gradation of the paper to relieve electric stress close to the conductor (diagrammatic)

inside the zone close to the conductor, which is subject to the strongest electrical stress, by increasing the permittivity to a limited extent (Figure 4.2) and thereby improve break-down strength without, at the same time, noticeably increasing the dielectric loss coefficient $\varepsilon_r \cdot \tan\delta$ of the insulating layer as a whole.

A further measure that was addressed in Section 3.4.2 for increasing electrical ratings while simultaneously reducing the loss coefficient was the possibility of using a sandwich arrangement of paper and polypropylene (PPL) as the solid component of the impregnated dielectric instead of conventional cellulose paper. The markedly reduced losses developed within this three-component insulation (paper – PP – impregnating medium), in particular, allows higher electrical stress without the risk of thermal break-down, making PPL dielectrics, together with XLPE, a prime choice for the range of the highest system voltages (500 kV to 765 kV) in present-day new installations.

As well as the purely electrical properties considered so far, a range of mechanical and material-related characteristics also play a role in the selection of certain types of paper for high-voltage cable technology:

- *tear length, breaking elongation, number of double folds* as a measure of mechanical resistance to stresses during processing (lapping at a certain pre-tension etc.),

- *absorbency and air permeability* as criteria for impregnation capability and finally

- the so-called *ash content* which is used to assess paper purity.

Table 4.1 summarizes the most important non-electrical characteristics of cable papers for the high-voltage and extra high-voltage range.

Table 4.1
Non-electrical properties of cable papers for high voltage and extra high voltage

Property	Dimension	Value
Paper web width	mm	15–40
Tape thickness	μm	80–200
Bulk density	g/cm^3	0,8–1,2
Porosity	cm^2	<4
Tear length, longitudinal[1]	m	>8000
Tear length, transverse[1]	m	>3500
Breaking elongation, longitudinal	%	>2
Breaking elongation, transverse	%	>4
Number of double folds	–	>1500
Absorbency (10 min)	mm	>10
Ash content	% by weight	<0,6
Air permeability	cm^3/min	<5

[1] Theroretical length of a paper strip at which this tears under its own weight when suspended at one end

Impregnating media

The main task of the impregnating liquids in the laminated paper dielectric is to replace the air and moisture content of up to 50 percent by volume, which is initially present according to Equation 3.25 between the individual layers of paper and within the paper, with an electrically high-grade, low-loss medium. To achieve this, impregnating media are used which are liquid at the temperatures $\geq 90\,°C$ prevailing during processing, but which exhibit varying *viscosity* under operating conditions depending on the type of cable:

- low-viscosity (liquid) impregnating media primarily for low-pressure oil-filled cables including types with polypropylene-paper laminate PPL,

- high-viscosity i.e. semi-liquid compounds for external gas-pressure cables and

- solid, so-called *non-draining compounds* for internal gas-pressure cables and high-pressure oil-filled cables in steel pipe.

The principal requirements in respect of impregnating media are summarized in Table 4.2.

The impregnating media consist in all cases of liquid hydrocarbons, which are either obtained from mineral oil by special refining or manufactured synthetically. With regard to low-viscosity impregnating media, some manufacturers continue to prefer mineral oil products, while semi-liquid compounds and non-draining compounds now consist almost exclusively of polymers of butene or isobutylene (*polybutene* or *polyisobutylene PIB*).

Synthetic polymers have superseded the mineral-oil-based impregnating compounds previously used, which had 20 to 25 % resin added to them to increase viscosity, for several reasons [57]:

- Their dissipation factor is markedly lower than that of mineral oil with resin additive.

- They exhibit outstanding resistance to ageing.

- Their viscosity can be adjusted precisely within virtually any limits by control of the polymerization process and then remains practically unchanged over the entire life of the cable.

- Their pour point is around or even below $-10\,°C$, meaning that even high-viscosity compounds remain fluid down to very low temperatures.

Table 4.2 Requirements for impregnating media

Category	Requirements
Electrical	• high breakdown strength • low dielectric dissipation factor
Physical	• good flow characteristics and impregnating capacity at processing temperature • low change in viscosity in the operating temperature range • low pour point in the case of low-viscosity oils
Chemical	• good oxidation resistance and ageing stability • chemical compatibility with other cable materials • stable absorption capacity in relation to gaseous decomposition products (especially hydrogen) from local electrical discharges

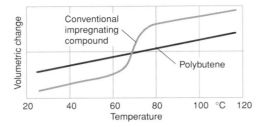

Figure 4.3
Volume contraction of impregnating compounds during cooling from processing temperature to ambient temperature (diagrammatic)

- They are distinguished by a small *cubic coefficient of thermal expansion* which is roughly constant over the whole temperature range of interest, whereas in the case of conventional compounds this characteristic passes through a pronounced maximum below the so-called solidification temperature at around 70 °C. As a consequence, this results in an almost abrupt decrease in volume when conventional compounds cool from the processing temperature to ambient temperature in the semi-liquid state (Figure 4.3), combined with a risk of shrinkage cavities forming in an originally saturated and cavity-free impregnated cable dielectric [58].

- Synthetic compounds can also be modified by adding synthetic, so-called microcrystalline wax to produce *non-draining compounds* such as are required for impregnating internal gas-pressure cables (for non-draining cables also at medium voltage). These non-draining compounds are also characterized by the advantageous properties described.

Even liquid cable oils are now upgraded by the addition of substantial amounts of synthetic hydrocarbons in the form of *alkyl benzenes* or their derivative *dodecyl benzene* DDB, or even completely replaced by these [57]. DDB unites two beneficial properties that cannot easily be combined in conventional paraffin or napthene based mineral oil with a few percent by volume of avid aromatic compounds to improve *gas receptivity*: it is active enough to take up decomposition gases (predominantly hydrogen H2) arising in local discharges, but in contrast to "natural" aromatic compounds which likewise take up gas it is largely homopolar (i.e. low-loss), chemically stable and resistant to ageing.

The chemical stability of DDB also makes it suitable as an impregnating medium for polypropylene-paper-laminates, which would swell in mineral oil; its outstanding gas receptivity is the reason why even the impregnating media for conventional low-pressure oil-filled cables are now generally enriched by adding a DDB proportion of at least 20 percent by volume.

4.1.2 Pretreatment of Materials Used

It has already been mentioned several times that paper and impregnating media have to be suitably treated prior to use as a high-voltage cable dielectric to achieve a satisfactory insulation capability. Moisture in particular has an adverse effect on both the dielectric and electrical properties of the dual-material system by increasing the dissipation factor and reducing breakdown strength (see Section 3.2.2, Figure 3.15). Moisture therefore has to be removed carefully from the components of the dielectric; the same also applies to the air inside the pores of the paper (Figure 3.4), which would prevent saturated impregnation with electrically high-grade impregnating media.

The moisture absorption of a material under given climatic conditions is governed by its sorption characteristics and can be seen from the sorption isotherms. The curves of these are determined by Henry's Law, according to which proportionality prevails between the concentration c of a gas or vapor in a substance and the related gas or vapor partial pressure p_p [59]:

$$c = K(\vartheta) \cdot p_p \tag{4.1}$$

The proportionality factor $K(\vartheta)$ is a material-specific, temperature-dependent constant, and the concentration c represents a weight-related variable:

$$\text{Concentration } c = \frac{\text{Weight of gas or vapor absorbed}}{\text{Weight of absorbing medium}} \tag{4.2}$$

It is sometimes useful in connection with water absorption to use the relative humidity w_r in the ambient atmosphere instead of the partial vapor pressure p_p as there is, in turn, a proportional relationship between the *relative humidity* w_r and the partial pressure at a constant temperature [60]:

$$w_r = p_p / p_{ps}(\vartheta) \tag{4.3}$$

where $p_{ps}(\vartheta)$ is the temperature-dependent *saturation vapor pressure* of water.

The term *water content w* is also used in preference to the term concentration c. Like c, this is defined by the relationship 4.2. Thus, Equation 4.1 can also be written in the form:

$$w = K(\vartheta) \cdot p_{ps}(\vartheta) \cdot w_r = S \cdot w_r \tag{4.4}$$

Here the factor S is described as the *water vapor solubility coefficient*. It is largely independent of temperature, at least in the parameter range of interest for technical drying

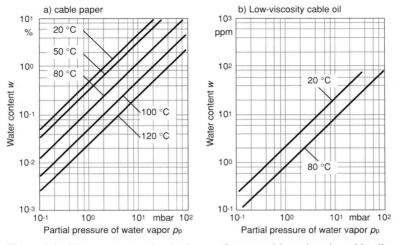

Figure 4.4 Water vapor sorption isotherms of paper and low-viscosity cable oil

95

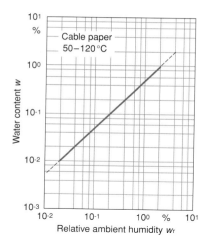

Figure 4.5
Water content of paper depending on relative ambient humidity [60]

processes between roughly 50 °C and 120 °C and moisture contents below the *condensation limit*, so that there is a proportional relationship between the water content in the medium in question and the relative humidity of its environment, regardless of the temperature [60, 61].

Figures 4.4 and 4.5 show examples of this. Figure 4.4 shows the water vapor sorption isotherms of cable paper and low-viscosity impregnating media based on mineral oil and Figure 4.5 shows the relationship between water content in the paper and the relative ambient humidity. Straight lines at an angle of 45° result in all cases which, in the case of the double logarithmic scale of the diagrams selected here, confirms proportionality both according to Henry's Law and Equation 4.4. This means that an effective reduction in the vapor concentration in the insulating material requires a corresponding drop in the related partial vapor pressure. Vacuum processes are therefore used in principle for drying and degassing: the relative ambient humidity can be used as a measure of the drying state.

Admittedly, considerable differences appear in the amounts of moisture which the two materials store under identical ambient conditions: the dimension "%" in the case of paper is in contrast to just a few ppm (parts per million = g of moisture per tonne of oil) in the case of insulating oil. The drying parameters that must be observed in the course of the vacuum treatment of paper or impregnating medium are therefore also different. Added to this is the fact that cable papers for use with high and extra-high-voltage are dried in two separate operations at different stages of the manufacturing process, while oil drying and degassing – mostly in association with a physical-chemical cleaning process – can be effected in just one step, *impregnating medium preparation*. Paper drying and impregnating medium preparation are therefore described below separately.

Paper drying

On the basis of its sorption isotherms (Figure 4.4) paper absorbs a water content of virtually 8 % of its weight when stored in normal ambient conditions of around 20 °C and

50–60% relative air humidity, corresponding to a water partial vapor pressure of 12 to 14 mbar. This water, which is predominantly condensed in the cellulose pores or adsorbed on the surface and of which only small amounts are chemically combined, not only impairs the insulating properties of the paper but also occupies a significant volume of the paper tapes. If these are applied to the conductor without further pretreatment as multi-layer insulation and only then have the water removed from them, the volume reduction resulting from this causes a perceptible break-up of the wound structure. Two undesirable consequences would be associated with this:

- So-called *soft spots* occur when the cable is bent, culminating in the formation of paper creases.

- The volume of solid which is lacking is filled in the course of the subsequent impregnation process with impregnating medium with a smaller electrical load carrying capacity.

Both effects cause a reduction in the resulting dielectric strength, particularly in relation to impulse voltage stress [26]. For these reasons a large amount of the moisture contained in the paper on delivery must be removed prior to its processing into cable insulation. To do this, the paper is first stored in dry conditions at a relative air humidity of around 3% (partial water vapor pressure around 1 mbar) and then lapped onto the conductor under the same conditions. A moisture content of $\leq 1\%$ is obtained according to the sorption isotherms (Figure 4.4) in the paper, further reduction of which content in the course of final drying following application of the insulation no longer gives rise to risk of break-up of the laminated structure.

The aim of this second phase of the drying process to which the so-called cable core assembly, i.e. the conductor completely lapped with paper insulation, is subjected, is a residual moisture content of less than 0.1% in the paper. This limit is partly derived from the dissipation factor curve of the impregnated cable dielectric in Figure 3.15 (Section 3.2.2), according to which the tan δ begins to rise measurably above around 0.1% moisture content.

Here the stated limit value must be achieved in all paper layers of the insulation, and thus in particular in those directly above the conductor. This means that the moisture must diffuse from the inner layers of the dielectric to its surface, which requires the presence of a concentration or partial water vapor pressure gradient from inside to outside [25]. A suitable means of creating an effective pressure gradient is evacuation of the drying tank into which the cable core assembly is inserted following application of the paper insulation (see Section 4.1.4). The required tank vacuum towards the end of the drying process is between roughly 10^{-2} and 10^{-3} mbar.

Higher temperatures promote paper drying in two respects:

- According to the sorption isotherms (Figure 4.4), the paper moisture falls dramatically as the temperature increases if the partial water vapor pressure in the environment is constant, i.e. drying is more effective, and

- The diffusion process is accelerated as the temperature rises, and the duration of the process is thus shortened.

Nevertheless, limits are imposed on too radical a temperature rise due to the fact that – even in a vacuum – the paper begins to decompose at an accelerated rate above approx-

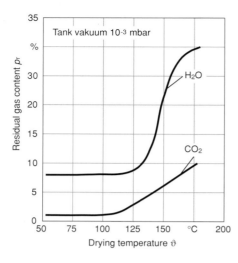

Figure 4.6
H_2O and CO_2 content in drying atmosphere of cable paper at different temperatures [25]

imately 125 °C. This process is characterized partly by an increase in the CO_2 content and water content in the residual gas composition in the drying vessel, as has been demonstrated by mass-spectrometric analysis [25]. As an example of this, Figure 4.6 shows the amounts of H_2O and CO_2 in the residual gas on reaching the steady state at the end of drying at different temperatures.

The moisture is for the most part originally chemically combined, so-called *water of crystallization*, the decomposition of which impairs mainly the mechanical properties of the paper and thus gives rise to the danger of embrittlement to the point of tearing under heavy mechanical loading (e.g. when the cable is bent). The decomposition of chemically combined moisture should therefore be avoided at all costs. The curve of the CO_2 content in the drying atmosphere indicates incipient decomposition reactions from as low as 110 °C, although on the basis of electrical and mechanical testing of the paper properties these have not proved critical. The optimum drying conditions are thus produced at around 120 °C and a tank vacuum of 10^{-2} to 10^{-3} mbar.

Apart from the parameters of tank pressure and temperature already considered, the duration of the drying process depends on a range of other parameters. These include the degree of pre-drying at the start of the vacuum treatment, the porosity (bulk density) and layer thickness of the paper, the width of the paper tapes and butt gaps and the lapping tension when applying the paper lapping as a measure of the strength of the lapping structure. The most important influencing parameter is the thickness of the cable insulation, which is determined in turn by the rated voltage of the cable. On first approximation, it has a quadratic relationship to the required drying period. Thus, periods ranging from a few days to roughly three weeks are required for high-voltage and extra high-voltage cables. Further details of the vacuum drying process are described below in connection with cable manufacture (Section 4.1.4).

Impregnating medium preparation

Although, as already stressed with reference to Figure 4.4, the absolute quantities of absorbed moisture are many times smaller under identical ambient conditions for insu-

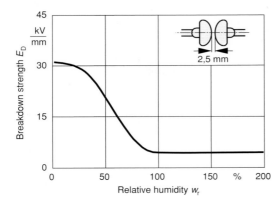

Figure 4.7
50 Hz breakdown strength of low-viscosity mineral oil in quasi-homogeneous field depending on relative humidity [62]

lating liquids than for paper, the electrical properties of the liquid almost react with even greater sensitivity to water than is the case with paper. As an example, Figure 4.7 shows the connection between the breakdown strength (50 Hz) of a low-viscosity insulating oil and its relative humidity. Starting from at least 30 kV/mm for the optimally dried insulating liquid, the breakdown strength falls as humidification increases to around 5 kV/mm, thus to just one-sixth of the initial value.

In addition, it is evident from this diagram that, even after optimum pretreatment, insulating oil has a considerably lower electrical strength at 30 kV/mm than impregnated paper with almost 70 kV/mm as shown in Figure 3.15 (Section 3.2.2). In conjunction with the field strength in the impregnating medium intermediate layers, which is virtually three times higher than the stress in the cellulose as a result of differences in the permittivity (Section 3.1.1, equation 3.1), this means that cable breakdown starts in principle in the impregnating medium. Particular importance is therefore attached to careful drying of the liquid. Here the actual drying and degassing process is associated with physical-chemical cleaning (deionization) of the liquid by means of fuller's earth mostly during the so-called *impregnating medium preparation*, leading to an additional reduction in conductivity and dielectric losses.

Water can be distributed in organic liquids in dissolved form or as more or less fine drops, the transition between the two forms of storage marking the *water vapor saturation solubility* or *water saturation solubility*. Unlike the situation in paper which, owing to its porosity and the capillary effect caused by this, has solubility characteristics diverging from those of homogeneous materials, the water saturation solubility w_L of liquids increases exponentially as the temperature rises:

$$w_L = w_{L0} \cdot \exp(-H/T) \tag{4.5}$$

where

$w_{L0,}\, H$ = material-specific constants
T = absolute temperature

This means that, as the temperature rises, impregnating media can store more moisture without this being precipitated in the form of drops. Nevertheless, an increase in the temperature favors the drying and degassing process – again carried out in a vacuum – for liquids too, the following reasons being responsible for this:

- The gas or vapor pressure within the liquid increases (greater partial pressure gradient in relation to the environment)

- The viscosity is reduced; the improved flow characteristics favor the formation of thin layers of liquid (shortening of the diffusion paths).

- The diffusion processes themselves are accelerated.

To make the diffusion paths for the gases and vapors to be removed as short as possible and at the same time offer the largest possible surfaces for the transfer, the liquid to be degassed is passed through *packed columns*, in which it flows via cylindrical tube sections, known as *Raschig rings*. Two-stage systems of glass or high-grade steel are standard in combination with series-connected fuller's earth filters, as shown in Figure 4.8:

- Ions and other impurities are removed from the liquid in the fuller's earth layer.

- If not already stored in heated storage tanks, the liquid is heated in the first column stage and evacuated to a vacuum of approx. 1 to 10 mbar, the mass of highly volatile gas and vapor components being vaporized first.

- In the second stage, a total pressure of 10^{-1} to 10^{-2} mbar (fine vacuum) prevails, sufficient for final drying to a residual moisture content of approx. 2–3 ppm.

As with paper, the lower the final vacuum and the higher the temperature selected, the quicker and more effectively the drying and degassing process progresses. However, limits are also imposed here on too effective an acceleration of this. As the temperature increases and/or the pressure drops, the components of the insulating liquid with a low boiling point are themselves inclined to vaporize, which can have a detrimental effect on important features of the impregnating medium. Low-viscosity liquids contain larger

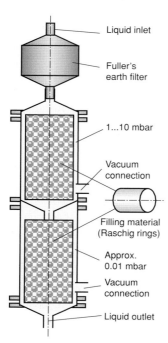

Figure 4.8
Two-stage packed column with bleaching earth filter connected on incoming side for deionization, drying and degassing of impregnating liquids (principle)

amounts of highly volatile components than those with a high viscosity, so that the optimum drying temperature varies depending on the type of impregnating medium:

- Room temperature up to 50 °C for low-viscosity oils and

- Approx. 90 °C for high-viscosity compounds,

- Final vacuum in each case 10^{-1} to 10^{-2} mbar as already mentioned.

Unlike low-viscosity oils, semi-liquid impregnating media must also be kept constantly at a higher temperature if they are to be sufficiently fluid. The separate heating on entry to the vacuum column mentioned above can therefore be omitted for these substances.

Impregnating medium preparation as a whole is considerably less time-consuming than paper drying. For this reason it generally takes place during manufacture at the same time as the impregnating process (see next section).

4.1.3 Electrical and Dielectric Properties

Satisfactory materials and production, including the requisite drying and degassing processes, are assumed in the list of selected electric and dielectric properties of paper cables in Table 4.3 below. Some of these form the subject of later chapters of this book (Section 4.1.4). In the interests of a unified overview, the summary also contains some characteristic quantities which have already been discussed elsewhere (Section 3.2.2, Figure 3.15, Section 3.4.2, Tables 3.6 and 3.7, Figs. 3.25 and 3.26 and Section 3.7.2, Figure 3.43). Finally, ranges are specified for the majority of attributes which cover the entire spectrum of high-voltage and extra high-voltage cables produced. This means that they encompass data of internal and external gas-pressure cables as well as those of low- and high-pressure oil-filled cables, as the differences in basic characteristics occurring between these types are relatively slight. If necessary these are highlighted and explained in the subsequent text.

One factor which is critical for the dimensioning of paper-insulated cables and thus for their insulation thickness in accordance with the remarks in Section 3.3.1 is the impulse

Table 4.3
Electrical and dielectric properties of paper-insulated high voltage and extra high voltage cables

Property	Test conditions	Value
Dielectric dissipation factor $\tan\delta$	50 Hz/20 °C	$1.5\ldots3\cdot10^{-3}$
Permittivity ε_r	50 Hz/20 °C	$3.5\ldots3.8$
Dielectric loss coefficient $\varepsilon_r\cdot\tan\delta$	50 Hz/20 °C	$5\ldots12\cdot10^{-3}$
Specific insulation resistance ρ	20 °C/1 min	$10^{16}\ldots10^{18}$ Ωcm
Specific insulation resistance ρ	90 °C/1 min	$10^{13}\ldots10^{15}$ Ωcm
Impulse voltage strength at cond. E_{DSmax}	20 °C/1…5/50 µs	$100\ldots130$ kV/mm
Impulse voltage strength at cond. E_{DSmax}	90 °C/1…5/50 µs	$70\ldots100$ kV/mm
Average short-term dielectric strength E_D	50 Hz/20 °C	$45\ldots70$ kV/mm
Average long-term strength E_H	50 Hz/20 °C	$25\ldots50$ kV/mm
Operating field strength at inner cond. E_{max}	50…60 Hz	$11\ldots16$ kV/mm

voltage strength at the inner conductor, which must always be sufficient to withstand the impulse voltage stress which may occur in operation and are consequently applied during type testing (Section 8.2.2). According to Table 4.3, this attains values of between 70 and 130 kV/mm depending on the temperature and cable design.

Thus, results in the upper range of the scatter band not only call for careful pretreatment and processing of the material, but also for the insulation to be constructed in a special manner. As already explained in Section 4.1.1, the use of highly dense papers of a small layer thickness close to the conductor makes it possible to reduce electrical stress specifically in this area, which is most heavily loaded owing to the cylindrical geometry (Figure 4.2) and to increase the strength at the same time. It has been possible to demonstrate the effect of this measure experimentally in step impulse voltage tests to breakdown at room temperature on short lengths of 400 kV low-pressure oil-filled cable [26]. Figure 4.9 illustrates the results in the form of two Weibull distributions. Although, at 104 and 123 kV/mm, the characteristic rated value in both cases is above 100 kV/mm, in the case of paper gradation every single test result markedly exceeds the 100 kV/mm mark, which is often taken as the characteristic dimensioning value (see Section 3.3.1).

An even broader band covers the values of AC voltage strength, as this is influenced not only by product-specific features, but also by the duration of electric stress. Although this effect is by no means as strongly pronounced in the case of multi-layer paper insulation as for thick polymer dielectrics (see Figure 3.13 and Section 4.2.1), even breakdown of paper cable is initiated, in the case of alternating voltage, by processes of destruction that are accelerated as the degree of loading increases. Thus, complete breakdown when a certain limit stress is exceeded is preceded by ionization and discharge processes within the intermediate layers of impregnating medium lying close to the conductor. These first age the impregnating medium, with heat being generated and then, as time progresses, also erode the adjacent layers of paper and finally create a discharge

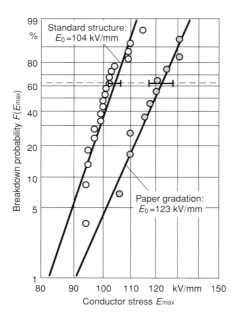

Figure 4.9
Impulse voltage strength of 400 kV low-pressure oil-filled cables with different paper gradation at ambient temperature [26]

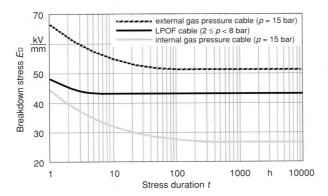

Figure 4.10
Life characteristics of various high-voltage cables with an impregnated paper dielectric

path through the insulating wall as a whole [63]. The absolute value of this limit stress and the steepness of the life characteristics which set in when this value is exceeded depend primarily on the type of impregnation as well as on the cable geometry and paper layer thickness, as clarified by Figure 4.10.

If careful preparation is assumed, the material-specific breakdown field strengths of the impregnating media under consideration do not differ significantly, and the permittivity values are also virtually identical at around 2.2, i.e. the same electrical load prevails within the intermediate impregnating medium layers in all three cable types evaluated here. The differences in breakdown time characteristics that can be seen in Figure 4.10 depend rather on the flow capacity (viscosity) of the impregnating medium, the homogenity of the intermediate impregnating medium layers and the pressure conditions in the insulation:

- In the case of gas-free impregnation of the dielectric, the inception field strength for ionization processes increases with the pressure acting on the impregnating medium. As a result, external gas-pressure cables have a higher load rating at around 15 bar than low-pressure oil-filled cables at operating pressures of between roughly 2 and 8 bar.

- Consequently, oil-impregnated cables exhibit more stable life characteristics above the limit stress, as their low-viscosity impregnating medium flows more swiftly out of areas with local pre-discharges and consequently ages more slowly than the semi-liquid impregnating media of gas-pressure cables.

- The load-time characteristics of internal gas-pressure cables, which are on the whole rather less favorable, are based on the fact that gas penetrates the insulation in this case, favoring the ignition conditions for ionization processes compared with gas-free dielectrics.

The classical dielectric properties of the insulation – dissipation factor, permittivity and also insulation resistance – can be improved considerably compared with the values listed in Table 4.3 for conventional paper dielectrics if laminates of paper and polypropylene are used instead of normal cellulose paper. Details regarding this and the options resulting from this for increasing the operating field strength and maximum transmission voltage were discussed extensively in Section 3.4.2 (Table 3.7, Figs. 3.27, 3.28) and thus require no further discussion here.

4.1.4 Methods of Manufacture

Regardless of the type of cable, the manufacture of paper-insulated cables can be broken down roughly into three stages, which will be described below in the order of their occurrence in the production process:

- Application of paper insulation (lapping process),

- Drying of the paper and impregnation with impregnating media
  (pretreated if applicable),

- Sheathing of the core with a metal sheath including any corrosion protection
  that may be required.

Parts of these production stages have already formed the subject of previous chapters and thus only minor additions are required here, e.g. paper drying or impregnating medium preparation.

Applying the paper insulation

The multi-layer paper insulation is applied to the conductor in a single operation on automated lapping machines, on which one roll of paper (i.e. with a diameter of ≥ 50 cm and of the width of the tapes to be processed) must be present for each insulating layer. The number of paper layers required is dependent in the first instance on the rated voltage of the cable to be produced, although the actual number of layers is not a standardized quantity but is determined by the respective cable manufacturer according to the manufacturer's own design criteria and operating experience. Reference values are around 100 paper layers for 110 kV and 200 to 250 layers for 400 to 500 kV.

Depending on the design of the installation, a so-called *winder* (a paper roll carrier rotating around the conductor) can take up to 12 individual rolls. Several of these winders form a complete *insulating machine*. Its overall length depends on the number of winders, this again being determined by the rated voltage of the cable to be insulated or the paper layers required for safe management of the operating and test stresses. If an insulating machine consists for example of n winders, which can each take m rolls, the maximum number of layers z to be applied by this installation amounts to

$$z = n \cdot m. \tag{4.6}$$

Extra high-voltage cables with the aforementioned 200 to 250 layers of paper therefore require installations comprising at least 20 winders with 10 to 12 rolls each, which thus attain a length of roughly 50 m. The direction of rotation of adjacent winders is always opposed, i.e. after every 12th layer of paper at most there is a reversal of lapping direction. The inner layers directly above the conductor are wound on in this process in principle opposite to the direction of twist of the outer conductor wires, to avoid "untwisting" of the conductor and so guarantee a mechanically strong cable core construction.

Figure 4.11 provides an impression of the construction of an insulating machine of the kind discussed which, to maintain the paper moisture content reduced to $\leq 1\%$ by predrying (see Section 4.1.2), is operated completely in an air-conditioned room at a rela-

Figure 4.11
Insulating machine for high voltage and extra high voltage cables in an air-conditioned room
(Section) [26]

tive air humidity of around 3%. The conductor is supplied from a *pay-off reel* (conductor storage roll) and the completely wound core (cable core assembly) then carried onwards through suitable air locks on both sides of the climatic chamber. The finished core assembly is either taken – once again in a normal environment – directly to the *impregnating tank* or first wound onto the *impregnating basket*, a rotating support, and supplied on this basket to the next production stage, drying and impregnating.

A complete core insulating system for paper cables therefore consists of the following three individual components (see diagram in Figure 4.12):

- Conductor storage coil (pay-off reel),
- Insulating machine with several single winders in a climatic chamber, separated from parts of the installation outside the chamber by air locks,
- Rotating impregnation basket for impregnating tank as core carrier.

The operation of large insulating machines calls for extensive know-how and a high outlay on control engineering in order, on the one hand, to obtain the strongest possible yet sufficiently flexible paper insulation and, on the other, to avoid overstressing the papers mechanically during processing (paper tearing). Here, the necessary flexibility of the

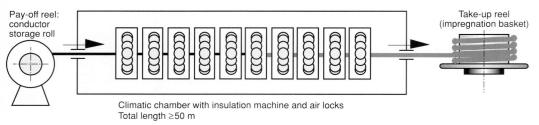

Pay-off reel: conductor storage roll

Take-up reel (impregnation basket)

Climatic chamber with insulation machine and air locks
Total length ≥50 m

Figure 4.12 Principal components for applying paper insulation

cable core assembly, which as yet is not impregnated, is particularly critical when winding it onto the core of the impregnating basket, as at this stage the impregnating liquid, which makes it easier for the paper tapes to slide over one another, is not present.

In contrast to the processing speed of extruders for synthetic cable manufacture (Section 4.2.3), that of winding machines is largely independent of the insulation thickness of the cables and is approx. 0.5 to 1m/min. The thickness is adjusted primarily by the number of paper layers chosen, which depends, according to Equation 4.6, on the number of active winders and paper rolls. Although fitting these influences the time taken to set up the machines, it does not affect the operating speed during the lapping process. Also, in contrast also to the methods of processing polymer material, production can be stopped at any time, e.g. to exchange empty paper rolls or to *join on* new conductors.

Drying and impregnation

As explained in Section 4.1.2, to obtain electrical properties which are satisfactory in the long term, the paper insulation must first be dried in a vacuum and then saturated with impregnating medium, which must likewise be dry and free of gas.

After placing the impregnating basket, if applicable, with the core wound onto it in a heatable vacuum tank (receptacle), the first stage in paper drying is the commencement of the evacuation process with the objective of creating a concentration gradient as a driving force for the diffusion of moisture from the cellulose pores by reducing the partial water vapor pressure outside the paper. To speed up diffusion, heating of the dry material to the maximum permitted temperature (approx. 120 °C, see Figure 4.6) is begun as soon as possible. High temperatures of this kind in the presence of air oxygen would trigger excessive paper ageing. Careful process control is therefore required to ensure that a sufficiently low tank vacuum exists prior to heating.

However, this gives rise to the problem of transporting the heat from the heated chamber walls to the dielectric: in a vacuum, the relatively ineffective *radiation* mechanism is virtually the only method available, while the considerably more efficient method of *convection* is largely ruled out owing to the lack of material with a flow capability in the recipient. Two alternative solutions are possible:

- Repeated interim venting with dry nitrogen as so-called interval drying, with the disadvantage that the evacuation process is artificially retarded;

- Direct *conductor heating* by DC. Advantage: heat is generated directly at the material being dried. Disadvantage: the risk of local excessive temperature, since different heat dissipation conditions are produced depending on the position in the wound-on cable core. However, conductor heating is the markedly more effective method, the temperature being monitored and controlled by measurement of the conductor resistance.

Even with very efficient pump sets, times ranging from several hours to days are required to produce the desired tank vacuum of 10^{-2} to 10^{-3} mbar, depending on the quantity and moisture content of dry material. It takes even longer to build up the required vacuum in the interior of the cable insulation, and the value of the tank vacuum is never equalled here owing to the pressure gradient existing across the insulation. As a rule of thumb, the quadratic equation already mentioned under 4.1.2

$$t \sim w^2 \tag{4.7}$$

applies to the relationship between the insulation thickness w and the required drying time t.

The following approximate recommended values thus result for the duration of the drying process:

- Approx. 2 days for 110 kV cores with a 10 mm wall thickness and

- Approx. 2 weeks for 380 kV cores with a 25 mm wall thickness.

Figure 4.13 [25] shows the development of the vacuum when drying paper cable cores of varying thickness. This diagram highlights once again the basic physical features of the process:

- The pressure in the tank drops considerably faster than in the interior of the cable, but the drop is also influenced by the insulation volume,

- The final vacuum in the interior is markedly poorer than in the receptacle and

- The duration of the process until a certain pressure value is achieved increases roughly with the square of the paper wall thickness.

When core drying is complete, the impregnating medium, which has been prepared in the meantime as described in Section 4.1.2 and heated, is conveyed under vacuum conditions into the drying tank and then, after a brief period, put under slight excess pres-

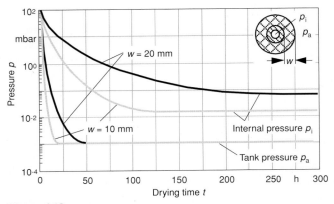

Figure 4.13
Pressure curves in the drying tank and in the hollow conductor of paper cable samples of varying insulation thickness during vacuum drying [25]

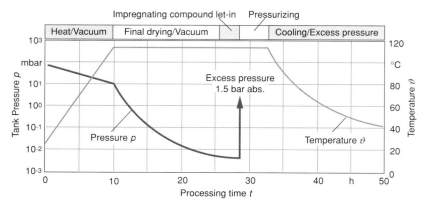

Figure 4.14
Curves of pressure and temperature in the impregnation tank during drying and impregnation of
the paper insulation ($w \approx 10\,\text{mm}$)

sure (0.5 bar) to promote penetration of the cellulose pores. Cooling of the impregnated
core to $\leq 40\,^{\circ}\text{C}$ then also takes place in this state, before the core is taken to the last stage
of the process, sheathing. This cooling is required to prevent gas and water vapor penetrating once more en route to the sheathing facility. It also serves to protect the dielectric
against overheating during production of the metal sheath, which is associated with substantial temperature loading (see next sub-section). The impregnating medium not
absorbed by the insulation in the drying tank is pumped off, reconditioned and kept for
further impregnation processes.

Temperature and pressure curves during the individual treatment stages for paper-insulated cable cores in the impregnating tank (drying and impregnation) are summarized
once more in Figure 4.14. The pressure curve in the initial stage during the heating phase
is particularly noteworthy here: despite operation of the pump set, the pressure does not
drop below approx. 10 mbar for a fairly long period, which is attributable to the
increased moisture release as a result of heating already mentioned above and the accompanying increase in vapor pressure within the paper.

In addition to flooding the chamber as a whole with impregnating liquid, it has proved
advantageous in two respects to supply the liquid oil for low-pressure oil-filled cables
with hollow conductors through the central oil channel of the conductor: the process
accelerates impregnation of the dielectric and also guarantees particularly saturated
impregnation of the paper layers close to the conductor which are the most highly electrically stressed.

Sheathing

The satisfactory electrical properties of the cable insulation achieved by paper drying
and impregnating medium degassing can only be maintained in the long term by an
absolutely gas- and vapor-tight sheath, which securely prevents the loss of impregnating
liquid as well as the renewed penetration of gases and vapor into the dielectric. Metal
materials continue to be used exclusively for this. If the individual cable cores are additionally under excess pressure in relation to their surroundings during operation, as is the

case with low-pressure oil-filled cables, the metal sheath must be capable of withstanding this stress also in the long term. In contrast to polymer-insulated cables, for which a thin metal foil provides adequate protection in principle against moisture, low-pressure oil-filled cables always require a solid metal sheath of lead or aluminum. The advantages and disadvantages of one design or another have already been considered in Section 2.3.4 and summarized there in Table 2.4.

In contrast to extruded cables, the sheathing of paper cables must follow on directly from core manufacture, primarily to avoid moisture absorption and impregnating medium losses. To avoid excessive thermal stress on the core insulation, the sheathing process is always executed continuously (without stopping), as the temperatures required to apply the metal sheaths are at least

- Approx. 250 °C when working with lead and

- Approx. 450 °C when using aluminum.

For the same reason, the metal sheath is cooled intensively with water immediately following application.

A differentiation is made between the two types of sheathing installation as follows:

- *Extruders* (can only be used for lead) and

- *Presses* (generally required for aluminum sheath manufacture).

Whereas lead extruders differ from corresponding machines for processing polymers only with regard to details such as processing temperatures and screw forms, presses are considerably more complicated installations. The most significant difference in relation to extruders is the type of material melting: purely thermal melting in the case of the extruder compared with thermal-mechanical deformation in the case of presses, using substantial, hydraulically generated forces (up to 40 000 kN = 4000 tonnes in large aluminum presses).

Owing to its ability to yield to thermal expansion of the cable insulation, lead can be extruded directly onto the core, similar to synthetic sheaths for XLPE cables. In contrast, the diameter of (corrugated) aluminum sheaths[1] must be markedly greater than the core diameter (see Figure 4.15). The overall diameter D is calculated here using the formula

$$D = d + 2s + 2t \tag{4.8}$$

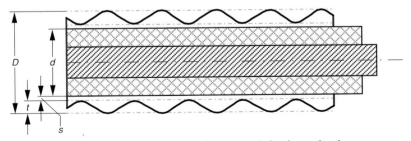

Figure 4.15 Determining the diameter of corrugated aluminum sheaths

---

[1] Corrugation is not required for core diameters of less than 50 mm.

where d = core diameter

s = expansion gap to avoid mechanical damage in the event of thermal expansion of the insulation under load

t = depth of corrugation troughs for a corrugated sheath

The expansion gap here must take account on the one hand of the – calculable – core expansion under full load and any necessary reserve for overloads in the event of a fault, yet on the other hand must not be too large, so as to ensure optimum heat dissipation from the cable. These considerations apply to an even greater extent to corrugated aluminum sheaths for XLPE cables, the dielectric of which has a considerably larger coefficient of thermal expansion than impregnated paper and does not create any impregnating medium layer between the core surface and the sheath to facilitate heat dissipation.

Finally, the metal sheath is provided with passive corrosion protection as described in Section 2.3.5, depending on the sheath material (lead or aluminum) and the ambient conditions to be expected when the cable is in operation. The simplest form is an extruded polymer sheath of PVC or HDPE; often further protective coatings of bitumen or bitumen-impregnated textile tapes are applied between the metal sheath and the polymer sheath.

4.1.5 Types of Cable

Paper-insulated high-voltage and extra high-voltage cable types can be divided into two main categories: *oil-filled cables* and *gas-pressure cables*. Further distinguishing features arise within the two groups, partly resulting from the operating pressure level and means of producing this, and also the type of impregnating medium. Another independent type consists of the group of *high-voltage DC transmission (HVDC) cables*. Reference has been made in previous sections to some aspects of these types of paper cable; for example, the type of thermal stabilization for oil-filled and gas-pressure cables has already been discussed in 3.1.2 and the field distribution under DC voltage stress in Chapter 3.7.

Since the basic electrical, dielectric and thermal properties of paper-insulated high-voltage and extra high-voltage cables do not differ significantly from one another, the choice of one type or another depends primarily on economic considerations in conjunction with local experience.

Oil-filled cables

Two different types of cable are grouped under the common term *oil-filled cables*. These are used for voltages up to 500 kV:

• *Low-pressure oil-filled cables or self-contained oil-filled cables*

Low-pressure oil-filled cables or simply "oil-filled cables", by far the most significant type of cable in Germany covering the entire rated voltage range from 60 kV to 400 kV and also 500 kV abroad. With a few exceptions, low-pressure oil-filled cables are now designed as a single-core cable with an armored lead or corrugated aluminum sheath (Figure 4.16). The impregnating medium is a partly or wholly synthetic, low-viscosity, electrically high-grade cable oil at an operating pressure of 2 to 8 bar absolute, which is maintained by sealed expansion tanks with nitrogen-filled diaphragms at intervals of a few hundred up to more than 1000 meters (see Figure 3.7). The hollow conductor which

Figure 4.16
400 kV low-pressure oil-filled cable
with corrugated aluminum sheath

is characteristic of low-pressure oil-filled cables is used as the connection between the expansion tanks and the dielectric (Figure 2.4).

Additional outlay is incurred on low-pressure oil-filled cables in cases where a section has to overcome fairly large differences in height in mountainous terrain. In cases such as this, the hydrostatic pressure p_s as a result of the oil column in the rising gradient is superimposed on the system pressure p_v which is predetermined by the expansion tanks and any dynamic pressure p_t triggered temporarily by a rise in temperature. At a density of $\rho = 0.9$ g/cm^3 of the cable oils which can be used, this hydrostatic element amounts to approx. 1 bar for every 11 m difference in height.

Since low-pressure oil-filled cables are rated for a maximum operating pressure of roughly 8 bar, fairly long graded sections have to be subdivided by the installation of so-called *stop joints* (see Section 6.3.1) into several sections separated hydrostatically from one another, at the lowest points of which respectively the total pressure

$$p_{ges} = p_v + p_t + p_s \tag{4.9}$$

may not exceed the value of 8 bar mentioned (see basic diagram in Figure 4.17). With $p_v \approx 2$ bar and $p_t \approx 1$ bar, stop joints are required at a height difference in excess of around 60m. In particular where cables are routed over extremely steep terrain, e.g. to connect underground hydro-electric power plants, a considerable requirement results in respect of accessories, which normally makes the choice of other cable types advisable for such uses.

- *Oilostatic cables or high-pressure oil-filled cables in steel pipe*
  Oilostatic cables or high-pressure oil-filled cables in steel pipe are a three-core cable variant without a metal sheath in a common, oil-filled steel pipe at an operating pressure of 15 to 18 bar, used predominantly in the USA, France and Japan up to 500 kV, and also occasionally in Germany (up to 110 kV). Since the cores have to do without any hermetically sealed outer sheath prior to insertion into the steel pipe, which is already in situ, the paper insulation generally has to be impregnated using a high-viscosity impregnating medium to avoid impregnating medium losses [64], although variants are in use abroad both with fluid impregnation ("*oil high-pressure cable*", Japan) and with non-dripping non-draining compound as the impregnating medium. Owing to the supply of pressurized oil in the steel pipe, the use of hollow conductors can be dispensed with for all high-pressure oil-filled cables; the core assembly is otherwise the same as for low-pressure oil-filled cables.

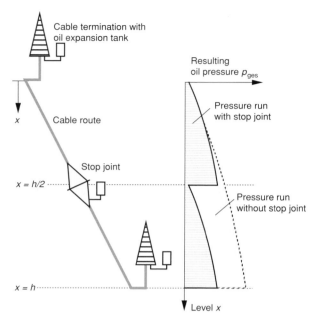

Figure 4.17   Pressure distribution in LPOF systems with stop joints on steeply graded routes

The high operating pressure necessitates the use of relatively expensive *pressure maintenance stations* with supply tanks, in which the oil is maintained in a vacuum, and automatically controlled pumps instead of the static compensating tanks with diaphragms, such as are sufficient for storing the impregnating medium for low-pressure oil-filled cables. Admittedly, this additional outlay can be partly offset by the fact that it is not necessary to subdivide a cable section hydrostatically by the use of stop joints, as described earlier, even over a fairly long gradient, since the system, which is rated overall for considerably higher compressive stress, will also tolerate greater hydrostatic oil pressures. Figure 4.18 shows an example of a high-pressure oil-filled cable for 110 kV.

In recent times, all oil-filled cables have increasingly become the subject of environmental debate, making their use in new installations for 110 kV virtually impossible, at least in Germany. The fear is that, in the event of cables being damaged, large-scale contamination of the soil by leaking impregnating medium will result. This is naturally the case to an

Figure 4.18
High-pressure oil-filled cable in steel pipe

even greater degree with regard to high-pressure oil-filled cables with their much greater volumes of liquid. These types of cable with their outstanding electrical properties have only succeeded in becoming established to a limited extent, not least for this reason.

Gas-pressure cables

Gas-pressure cables have a number of characteristics in common with the oilostatic cables described above: both variants are executed as three-phase cables, both are drawn through a common steel pipe and both are operated at a pressure of 15 bar to maintain thermal stability (Section 3.1.2). The conductor design also corresponds, being executed in each case without a hollow duct. The paper insulation, which is identical in principle with that of oil-filled cables of the same rated voltage, is basically impregnated with a high viscosity medium, if applicable with non-draining compound, which is now exclusively a polybutene-based, fully synthetic impregnating medium. The critical difference between oilostatic and gas-pressure cables consists in the type of pressure medium used: liquid oil in the case of oilostatic cables and nitrogen N_2 for gas-pressure cables.

The upper limit of the rated voltage for gas-pressure cables is 275 kV. In Germany, they are used up to $U_N = 110$ kV and today still account for a market share of around 50% at this voltage level. Besides their small line width, low *reduction factor* (see Section 7.3) and outstanding mechanical protection due to the common core sheathing in a steel pipe, the good acceptance of these cables is due to their ecological harmlessness, as there is no fear of any notable soil contamination by leaking impregnating medium in the event of damage to the cable on account of the high viscosity.

There are two sub-groups in the gas-pressure cable category: internal gas-pressure cables and external gas-pressure cables..

- *internal gas-pressure cables*
  With the exception of the pressure medium, internal gas-pressure cables (Figure 4.19) are virtually identical in construction to oilostatic cables. The three high-viscosity cores, impregnated for the most part with non-draining compound, in a common steel pipe and without a metal sheath are in direct contact with the surrounding nitrogen, so that the pressure gas penetrates into the dielectric and can, if applicable, stabilize electrically any voids (see Figure 3.8) which exist or may arise due to thermal load variations by increasing the ionization voltage. This stabilizing mechanism can be improved further by adding a certain percentage volume (roughly 20%) of the electronegative insulating gas *sulfur hexafluoride SF_6* to the nitrogen. An increase in the breakdown field strength of the gas to almost double the comparative value of pure nitrogen is thereby achieved without adversely affecting the other properties which are critical for its use as a pressure gas in the cable, such as its flow and condensation

Figure 4.19
Internal gas-pressure cable for 110 kV with common armoring as protection during drawing in

characteristics [65]. One positive side-effect of this measure is that any leaks in the system can be detected considerably more reliably by means of suitable leak detection devices calibrated for SF_6 than would be the case with pure nitrogen, the principal constituent of atmospheric air.

Nevertheless, the dielectric strength of internal gas-pressure cables remains at the lower limit of the spectrum of all paper-insulated high-voltage cables according to Table 4.3 (Section 4.1.3), as the presence of voids in the dielectric, even with the optimum gas filling, signifies a slight weakening compared with saturation-impregnated insulations. The non-drip properties of the impregnating medium prove advantageous when internal gas-pressure cables are used in mountainous terrain, where they can surmount any differences in altitude, even in the case of almost vertical line routing, without hydrostatic subdivision by means of stop joints.

- *External gas-pressure cables*
  External gas-pressure cables (Figure 4.20) are distinguished from the internal gas-pressure cables described above primarily by the presence of a gas-tight lead sheath over each individual core and the resulting freedom from gas of the dielectric. To avoid ionization processes as a prerequisite for thermal stability, the insulation is first impregnated completely free of voids ("saturated") and this state is maintained under all operating conditions by the compressive effect of the gas pressure on the core sheathing (see Figure 3.8 in Section 3.1.2). To support this characteristic of the dielectric, also described as "breathing", in thermal cycling, the core has a slightly oval contour resulting from appropriate shaping of the conductor (Figure 2.4), and the impregnating medium retains a residual flow capacity even at lower temperatures. This results in one disadvantage of external gas-pressure cables, which are otherwise outstanding in electrical terms: in locations on extremely steep slopes, with very large differences in height, impregnating medium may be displaced, as happens with oil-filled cables, and the lead sheath may expand under its own weight in low-lying sections of cable. However, the high operating pressure in conjunction with the semi-liquid character of the impregnating medium allows for markedly greater differences in height than is the case with low-pressure oil-filled cables. In addition, the expansion of the lead sheath can be countered using reinforcement helixes of steel or hard copper.

As indicated by the above comments, the high electrical ratings of gas-pressure cables, regardless of their individual construction, are partly attributable to the stabilizing effect of the pressurizing gas. Thus the question arises of how the insulation reacts in the event of a sudden loss of pressure, caused for example by mechanical damage.

For external gas-pressure cables, an event of this kind does not immediately signify any direct risk of breakdown, as at the time the damage is incurred the insulation is free of

Figure 4.20
External gas-pressure cable for 110 kV

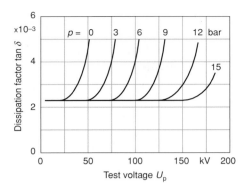

Figure 4.21
Voltage dependence of the dissipation factor of 110 kV internal gas-pressure cable at various equilibrium gas pressures (according to [66])

voids and is able to safely withstand the operating voltage applied under these conditions. The danger of increased lead sheath expansion and the resultant displacement of impregnating medium with the subsequent formation of voids in the dielectric, as shown in Figure 3.8, only arises due to repeated thermal load variation in the pressure-free state. Since fairly large load variations occur once per day at most in the high-voltage grid (see daily loading curve, Figure 3.29, Section 3.5.1), external gas-pressure cables can normally continue to be operated without pressure for at least 24 hours, and for the most part even 48 hours in today's essentially evenly loaded grid, before they have to be switched off.

From the physical point of view, a loss of pressure in internal gas-pressure cables can be regarded as critical, since their dielectric may have voids from the outset in which ionization processes are to be expected even at the rated voltage without a stabilizing gas filling. Appropriate tests on 110 kV internal gas-pressure cables in the 1960s have in fact revealed that ionization processes in the form of a pronounced increase in the dissipation factor can be detected well below the rated voltage if the equilibrium pressure is too low [66]. Figure 4.21 shows the relevant results.

Although the emphasis here is on the term *equilibrium pressure*, i.e. the pressure conditions which can be measured externally are present identically throughout the dielectric. Fairly long time in fact elapses following a loss of pressure before the gas has diffused from the voids in the interior of the insulation, especially those in the highly electrically stressed zones on the core surface, leaving behind ignitable voids. Most manufacturers therefore allow pressure-free operation at rated voltage for 24 to 48 hours even for internal gas-pressure cables. This time is generally sufficient to locate the damage and prepare for repairs, so that the actual down time of gas-pressure cables remains limited in the event of damage to the gas supply system to the time taken to carry out the repair.

Pressurized cable systems must, of course be monitored constantly, like oil-filled cables, to detect leaks promptly.

HVDC cables

The basic design of HVDC cables is essentially the same as that of high-voltage cables for three-phase AC operation and, as shown in Figure 2.2 (Section 2.2.2), comprises the elements conductor, insulation layer, field-smoothing layers, metal sheath, steel wire armoring and corrosion protection. However, the design of the individual structural elements is adapted to the specific requirements of direct voltage operation and is typically distinguished from conventional high-voltage or extra high-voltage cables as follows:

- The conductor design can be chosen without regard to current displacement effects. Stranded, compacted circular or shaped-wire conductors with a central oil duct if necessary are common. The material used is copper owing to its good specific DC conductivity. The cross-sections of HVDC cables produced up to now are 1200 to 1600 mm^2 to limit the critical overall weight for cable laying at sea, although a copper cross-section of 3000 mm^2 is envisaged for the Japanese submarine cable project described in [46] (Section 3.7.1).

- Up to now, multi-layer dielectrics of impregnated paper have been used exclusively as insulation (see Table 3.12). Both low-viscosity liquids such as are used in low-pressure oil-filled cables and semi-liquid compounds, as are commonly used to impregnate gas-pressure cables, can be used as an impregnating medium. Owing to the very low risk of ionization discharge in the case of DC voltage, it is even possible to dispense with pressurization and operate the cable as a normal maintenance-free paper-insulated, mass-impregnated cable, which proves advantageous with regard to long submarine cable links in particular. Here, operating field strengths of between 25 and 30 kV/mm are customary at the inner field-limiting layer, with field strengths of up to 40 kV/mm being possible for oil-filled cables under increased operating pressure (20 bar) [67, 68]. As a result, an insulation thickness of 17.5 mm is feasible, for example for a mass-impregnated cable for 400 kV DC [45], although, with a stress of this kind, the maximum conductor temperature of the cable is limited to 50 °C to counter too intense a field magnification in the outer insulating layers as per Figure 3.44 and, at the same time, counter the risk of increased void formation during thermal cycling of this construction which, in the sense of Figure 3.8, is not thermally stabilized.

- Like all cables with an impregnated multi-layer dielectric, HVDC cables also have a metallic sheath in principle, for which lead is preferred as a material for reasons of flexibility and the overall diameter. In submarine cables, the lead sheath is encompassed by expensive mechanical protection in the form of single- or multi-layer steel wire armoring, which stabilizes the cable during laying and against the water pressure and current on the sea bed. The corrosion protection has to be modified to suit the requirements of the cable environment (e.g. salt water).

DC cables are operated in two different system configurations: single pole and double pole. Single pole systems consist of a single cable with + 300 kV, for example, between the conductor and sheath. Return current conduction here takes place via the surrounding earth and/or sea water. In contrast, double pole systems have one cable for each of the outward and return currents, which are operated by analogy with the previous example at ± 300 kV and consequently attain double transmission capacity, although this advantage is balanced by the investment outlay compared with single pole systems, which is likewise precisely double. The maximum transmission capacity of HVDC cable lines installed to date is 600 MW. However, the Japanese project [46] already mentioned is aiming to attain new orders of magnitude in this respect: 1400 MW at ± 250 kV is planned in the first expansion stage and 2800 MW at ± 500 kV in the final stage.

4.2 Cables with Extruded XLPE Insulation

Cables with extruded solid insulation – basically produced as single-core cables – have already featured in several chapters of this book (Sections 2.2.3, 2.3.2, 3.4.2, 3.5... among others), in which as well as cross-linked polyethylene, the thermoplastic starting material for this, LDPE, and synthetic rubber, EPR, were looked at in this category. High-voltage and extra high-voltage cables with LDPE or EPR insulation have been used successfully in some countries for years; however, their present importance world-wide and all the more so in Germany falls far short of that of XLPE cables. The information provided under the heading "... high-voltage and extra high-voltage cables currently available" in this main section is therefore intentionally limited to XLPE.

4.2.1 Design of Single-Layer Dielectric and Properties of its Components

Insulating compounds

The basic compound polyethylene (PE) is a polymerization product from the *monomer* ethylene (C_2H_4). During *polymerization*, first the double bond of the monomer is broken, for example by supplying energy, or due to the effect of a catalyst and the radicals arising are then strung together in the form of a chain [69]. In this type of *polyreaction*, no reaction products of any kind such as water or low-molecular hydrocarbons are released, a basic prerequisite for the extraordinarily good dielectric properties of polyethylene (see Section 3.4.2, Figs. 3.23, 3.26, Table 3.7). Since, in addition, ethylene is structured entirely symmetrically – is *nonpolar* – and polymerization can be initiated by high pressure and high temperatures alone without foreign substances, the polymer PE ideally also has a completely regular structure without any charge imbalances, side chains or impurities (*linear PE*; Figure 4.22).

The individual chains are parallel in broad areas and thus form a macromolecule with its individual components ordered in some cases in a manner similar to crystallites, for which reason polyethylene and similarly structured *polyolefins* are described as *partial crystalline polymers*. However, at increased temperatures, this molecular microstructure becomes increasingly disturbed, and the crystallinity is reduced until the material enters a completely amorphous (glassy) and simultaneously fluid state as the so-called *crystallite melting range* is exceeded (Figure 4.23). Depending on the structure of the material, the softening temperature of PE has values of between roughly 110 and 130 °C. On re-cooling, polyethylene solidifies once more without any change in its chemical, morphological and physical properties and is therefore to be classed under the category of *thermoplastics*.

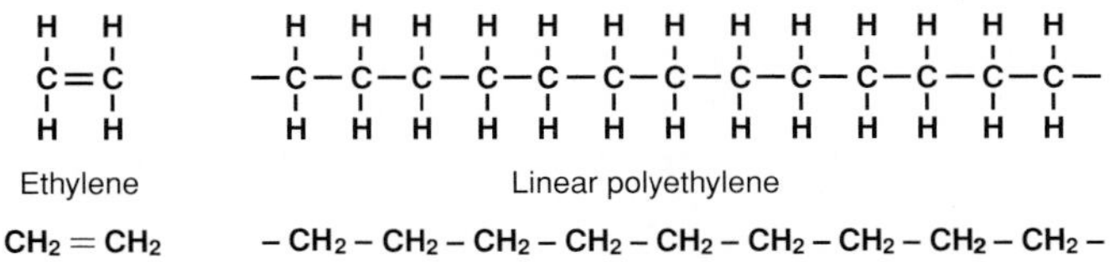

Figure 4.22
Structural formulae of ethylene and linear polyethylene in different forms of representation

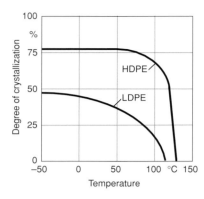

Figure 4.23
Temperature dependence of PE crystallinity
(according to [70])

During commercial polymerization at temperatures of up to 200 °C and pressures between 1500 and 3000 bar in the presence of oxygen, however, branches may form more or less frequently at individual C atoms as a result of radical initiation, and side chains of varying length are created (Figure 4.24). This chain branching at so-called *tertiary C atoms* inevitably leads to a loosening of the molecular structure and thus to a reduction in the density of the material, giving rise to the now customary term of *Low Density Polyethylene LDPE* for PE polymerized in a high-pressure process, instead of the name *high-pressure PE* which was common previously. The molecular microstructure mentioned above is also disturbed and the crystalline content decreases along with the upper limit of its thermal melting range, and as a consequence of this the maximum application temperature permitted is also reduced. The branched polyethylenes of low density used in cable technology have on average 1000 C atoms of a chain of 25 to 30 branching points; the resulting density is around ≤ 0.92 g/cm³, the crystalline proportion is 45 % maximum, the softening temperature reaches 110 to 115 °C and the permitted continuous operating temperature is 70 °C.

The less branched *High Density Polyethylene HDPE* also used previously in cable technology with a crystallinity of more than 70 % (Figure 4.23) only softens at just below 130 °C, therefore permitting a continuous use temperature of 80 °C. The large crystalline content makes the material very strong mechanically, but at the same time renders it difficult to work and makes it relatively hard and inflexible at lower temperatures, which is why it is now normally only used as a sheath material.

The relatively low crystallite melting temperature of LDPE in particular harbors a risk that in the event of damage, large lengths of the cable will become useless due to the thermal effect of the short-circuit current, because the heavy conductor is displaced eccentrically in the softening insulating sleeve. In conjunction with the low continuous

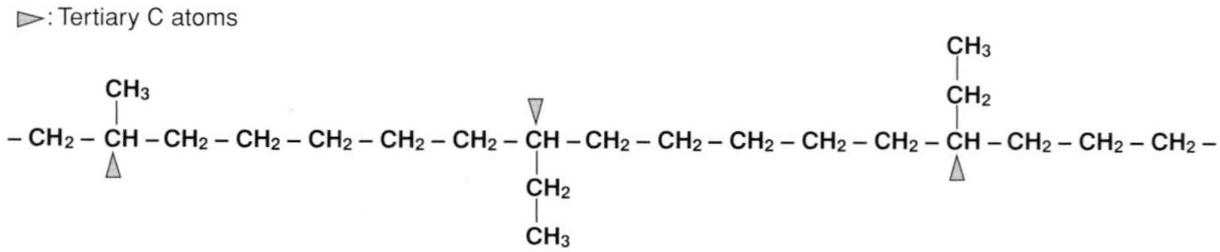

▷: Tertiary C atoms

$$- CH_2 - \underset{\underset{\triangle}{CH_3}}{CH} - CH_2 - CH_2 - CH_2 - CH_2 - CH_2 - \underset{\overset{\triangledown}{\underset{CH_3}{CH_2}}}{CH} - CH_2 - CH_2 - CH_2 - CH_2 - CH_2 - \underset{\underset{\triangle}{CH_2 \, / \, CH_3}}{CH} - CH_2 - CH_2 - CH_2 -$$

Figure 4.24   Structural formula of branched polyethylene

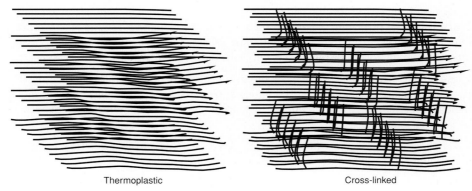

Thermoplastic   Cross-linked

Figure 4.25   Principle of cross-linking of PE chain molecules

operating temperature of thermoplastic polyethylene, this fear provided the impetus for the development of cable insulations of *cross-linked polyethylene XLPE*.

Whereas the molecule chains of thermoplastic PE slide against one another virtually unhindered on heating and can thus flow apart, this complete softening is prevented in the case of XLPE by the presence of individual cross-linkages between the main chains – the cross-linking (Figure 4.25). When the crystallite melting range is exceeded, the material therefore assumes a rubber-elastic rather than a fluid state, with finite tensile and compressive loading capability (Figure 4.26), for which reason XLPE is also classed in the *thermoelastic* group. With no change in its mechanical properties in the range of normal ambient conditions, XLPE can be stressed thermally to a considerably higher degree than thermoplastic PE; the continuous operating temperature rises to 90 °C, and the danger of conductor displacement is excluded, even where the crystallite melting range is exceeded markedly for a short time. In consequence, a short-circuit temperature of 250 °C instead of the 150 °C for LDPE is permitted in some standards for XLPE cables (see Table 3.2).

The cross-linking of polyethylene insulation of high-voltage and extra high-voltage cables is performed chemically by the use of a *peroxide*, a molecule with a characteris-

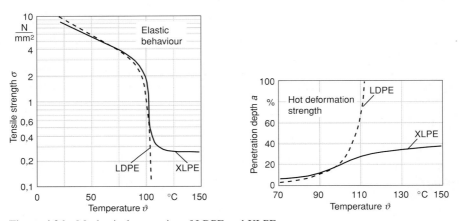

Figure 4.26   Mechanical properties of LDPE and XLPE

tic oxygen bond (R−O−O−R), and the application of heat [8]. The peroxide is generally contained in the insulating compound on delivery, like certain processing aids and ageing stabilizers (*antioxidants*), although some manufacturers prefer so-called *peroxide direct dosing*, in which the additive is only added during the production process.

Cross-linking takes place continuously during cable production regardless of the time the peroxide is added (Section 4.2.3) and basically comprises three reaction stages, as highlighted diagrammatically in Figure 4.27 taking the common method using *dicumyl peroxide* as an example:

- *Peroxide activation*, i.e. splitting of the peroxide into radicals with free valencies (including R−O• und CH$_3$•) by the addition of heat (approx. 200 °C);

- Formation of *polymer radicals* due to the release of single H atoms from PE chains and addition of these to R−OH or CH$_4$, forming free valencies in adjacent PE chains. This process commences preferably at the tertiary C atoms described earlier, because the linkage energy for atomic hydrogen at these points is slightly less than in the undisturbed areas of the chain molecules. The presence of branches thus makes cross-linking easier.

- *Cross-linking (curing)* by recombination of the PE radicals at the free valencies, forming a spatial network.

Apart from the cross-linking, these reactions also produce a range of low-molecular fission products, the composition of which depends partly on the choice of peroxide. Typical reaction products of *dicumyl peroxide*, which is widely used as an aid to cross-linking, are *cumyl alcohol* and *acetophenone* [71]. Methane CH$_4$ is also always released, a flammable hydrocarbon gas, which has to be carefully removed – *vaporized* – prior to further treat-

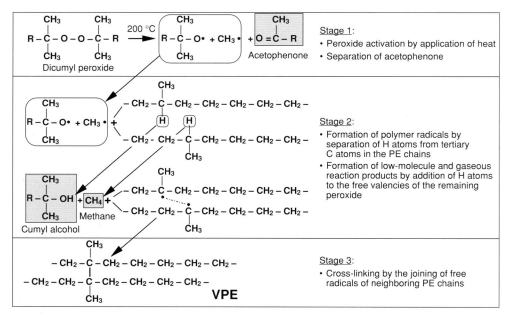

Figure 4.27 Stages in peroxide cross-linking of polyethylene

ment of the cable. Methane can also form bubbles in the still soft molten mass, in which partial discharges later occur under electrical loading. This danger must be countered by applying adequate excess pressure during cross-linking. Other details of the cross-linking process are covered in Section 4.2.3.

Owing to the additives and reaction products mentioned, which remain in some cases in the insulation, XLPE does not have quite such a regular structure as thermoplastic polyethylene with the idealized structural formulae of Figs. 4.22 and 4.24. Added to this are microscopic impurities resulting from the compound manufacture and cable production. All these foreign substances can influence the electrical and dielectric properties of XLPE insulations to a substantial degree, so that the characteristics of the extruded solid dielectric must always be assessed in relation to its level of purity (see also below).

Semi-conductive layers

The principal functions of the semi-conductive layers as highlighted in Section 2.3.3 are the homogeneization of local field magnifications e.g. at individual wires of the conductor or screen (Figure 2.6) and the avoidance of gap formation in the field area under thermo-mechanical loading. This presupposes a permanently strong bond between the semi-conductive layer and the insulation as well as adequate conductivity.

Semi-conductive layers for XLPE cables are therefore themselves based on cross-linkable polyethylene, which is extruded in a single operation with the insulation and welded permanently to the dielectric *prior* to the actual cross-linking (triple extrusion, see Section 4.2.3). The close chemical relationship of the base compounds ensures virtually identical thermo-mechanical properties of the insulation and semi-conductive layer, so that both structural elements extend under the influence of heat or contract on re-cooling to the same degree. In conjunction with the permanent bonding this ensures fully the required long-term freedom from gaps, even during stress cycles.

Since polyethylene, like other polymer materials, has not been available to date with sufficient intrinsic conductivity for use in cable technology, the required conductive capacity is achieved by mixing in suitable carbon blacks with the basic polymer. The resulting conductivity depends here primarily on the amount and to a limited extent on the type of carbon black [11, 72]. To maintain the minimum values stipulated in relevant standards, considerable proportions of carbon black of between roughly 35 and 40 percent by weight are required, the homogeneous integration of which is only achieved satisfactorily when *co-polymers* are used as the base compound. The term co-polymer is used to describe polymers composed of different monomers in contrast to polyethylene as shown in Figure 4.22 or 4.24.

The base compounds used for XLPE semi-conductive layers are normally so-called EEA (Ethylene-Ethyl-Acrylate) or EBA (Ethylene-Butyl-Acrylate) co-polymers, in which the conductive particles of carbon black settle adjacent to one another to form *agglomerates*. Microscopically thin insulating layers of polymer material remain between adjacent agglomerates, resulting in a material structure in the semi-conductive layers consisting essentially of two components, as illustrated diagrammatically by Figure 4.28 (the quasi-parallel alignment indicated here of the carbon black agglomerates results from the extrusion process during cable manufacture and causes a certain anisotropy of the properties of the semi-conductive layer).

121

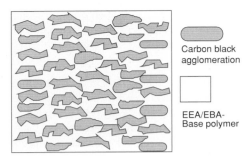

Carbon black agglomeration

EEA/EBA-Base polymer

Figure 4.28
Schematic structure of extruded semiconductive layers in XLPE cables (according to [11])

When dimensioning the addition of carbon black, an optimum has to be found between opposing influences as the carbon black content increases: conductivity on the one hand and mechanical and technological features such as homogeneity, elasticity, and extrudability on the other. High quantities of carbon black promote conductivity but impair the other material properties. Even slight deviations from the optimum can cause a marked loss of quality of the semi-conductive layer, which has a direct effect on the cable properties, as recent investigations show (Section 4.2.3) [72, 73]. The homogeneity of the boundary layers with the insulation in particular plays an increasing role as operating field strengths increase, which has prompted the development of special "*super smooth*" semi-conductive compounds for extra high-voltage cables [74]. Many semi-conductive compounds are also susceptible to moisture and therefore have to be dried out carefully prior to processing.

Apart from the base copolymer and carbon black, semi-conductive compounds contain a range of further constituents in a far smaller concentration: slip additives as an aid to processing, stabilizers and peroxides for chemical cross-linking. The amounts of the individual components are summarized in Table 4.4 and juxtaposed with the minimum requirements prescribed by standardization with regard to conductivity at various temperatures. The conductivities listed there apply in principle to the finished semi-conductive layer on the cable in its new state and following one week's so-called total-cable aging at 100 °C. All values have to be demonstrated experimentally in a likewise standardized procedure in the context of type testing (Section 8.2.2).

Table 4.4 Composition and minimum conductivities of semiconductive layers

Component	Content	Purpose
EEA/EBA copolymer	<60 % by weight	Base material permanently bonded to insulation
Carbon black	35–40 % by weight	To provide electrical conductivity
Stabilizers	<1 % by weight	Improvement of resistance to thermal-oxidation
Slip additives	<5 % by weight	Improvement of extrudability and homogeneity
Peroxides	<2 % by weight	Cross-linking agent
Requirements	specific d.c. resistance: • internal: max. 10^5 Ωcm at 90 °C and 130 °C • external: max. 5×10^4 Ωcm at 90 °C and 110 °C • before and after total-cable aging for 7 days at 100 °C	

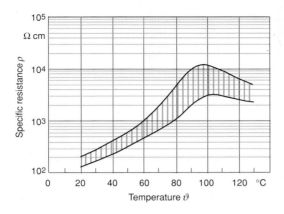

Figure 4.29
Temperature characteristic of specific
DC resistance of XLPE semi-conductive
layers for high-voltage cables

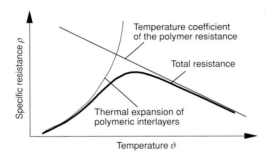

Figure 4.30
Reason for maximum in temperature curve
of the specific resistance of carbon-black
filled semi-conductive compounds for
XPLE cables (principle)

Characteristic features of carbon-black-filled XLPE semi-conductive layers are a maximum in the temperature curve of the specific DC resistance and considerable scattering of the results in spite of a composition which is identical in summary, as Figure 4.29 clarifies, taking the resistance band from several measurements on semi-conductive layers for high-voltage cables as an example. The lower the resistance at the maximum and the smaller the scatter, the higher the quality of the semi-conductive compound.

Both phenomena described earlier are related to the mix structure of the semi-conductive compound. Large dispersions result from the fact that even the slightest differences in the distribution of the individual particles of carbon black during extrusion, cross-linking and solidification readily influence the thickness of the microscopic intermediate polymer layers and thus the resistance. The reason for the temperature curve of the DC resistance is to be found in opposing mechanisms in the intermediate polymer layers (see Figure 4.30): in the lower temperature range, the resistance increases initially on heating, as the insulating layers between the conductive particles expand thermally; by contrast, at higher temperatures the negative temperature coefficient of these layers dominates, causing a reduction in the resistance of the configuration as a whole.

4.2.2 Electrical and Dielectric Properties

As already indicated in the preceding section, the electric and dielectric properties of complete XLPE cables are the result of interaction between the insulation and semi-conductive layers. Thus the dissipation factor of the cable (see Figure 3.23 and Table 3.7) is

123

slightly higher than that of an XLPE dielectric alone, as insulation and semi-conductive layers form a series connection and the semi-conductive layers have a finite resistance according to Figure 4.29. This is true to an even greater degree of cable breakdown which, due to space charges (Figure 3.3), is initiated in most cases at one of the two semi-conductive layers, most likely the inner layer owing to the cylindrical geometry of the dielectric [75].

Preparing XLPE cable technology for higher rated voltages therefore calls for improvements not only in the insulating compound, but also in the semi-conductive compound and the manufacturing technology used to process both materials to form a cable. One additional difficulty in the case of synthetic dielectrics is that relatively generous wall thicknesses were allocated when dimensioning the first high-voltage cables and it has only been possible in consequence to accumulate operating experience at low field strengths in the first instance (example: 110 kV XLPE cable with 18 mm insulation, corresponding to an average field strength of just under 3.6 kV/mm). Since such a low utilization of insulating material cannot easily be transferred to higher rated voltages for economic and manufacturing technology reasons, the introduction of each new voltage level has always been accompanied by an increase in the operating field strength (Figure 4.31) until, in the case of current 500 kV XLPE cables with an average field strength of roughly 10 kV/mm and a maximum stress of 16 kV/mm, values were finally arrived at which have already been common for many years in the case of paper-insulated cables.

It is characteristic of extruded polyolefin insulating materials that, in spite of these considerable stress increases in the course of recent cable development, a substantial gap continues to exist between the operating field strength and the *material-intrinsic breakdown strength* of the insulation measurable under optimum conditions of several 100 kV/mm [28, 76]. In addition, it is not normally possible to induce breakdown using alternating voltage on actual XLPE high-voltage cables or even extra high-voltage cables, so that it is necessary to rely on substitute tests in some form or other. These are either carried out on considerably thinner *model cables*, or forms of loading other than the continuous stress present in operation are applied by an impulse voltage [21, 22]. To achieve an approximate conversion to the actual operating conditions in the cable, suitable transformation formu-

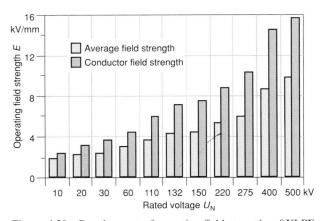

Figure 4.31   Development of operating field strengths of XLPE cables from 10 to 500 kV

lae are then applied – the volume transformation (equation 3.9) or the life law (equation 3.4), although the statistical dispersion which always exists has to be compensated for with relatively large safety deductions.

As an example, reference is made to the stress/time diagram ("life curve") in Figure 4.32, which has been invoked repeatedly by the authors of this book in a similar form elsewhere for dimensioning XLPE extra high-voltage cables [22, 30]. It is based on numerous life trials on XLPE high-voltage and extra high-voltage cables over an operating period of up to three years and leads theoretically to a life exponent N of roughly 12. Admittedly, only five of 34 long-term tests caused a breakdown; in all the other tests, especially all tests over an operating period of more than 30 hours, the cable specimens remained undamaged. Since the cables which broke down also were all taken without exception from the oldest production batches at the beginning of the 1980s, the life curve in Figure 4.32 says nothing more than that XLPE high-voltage cables are equal to the loads applied for this diagram and the life exponent has to amount to at least $N = 12$. However, it is not possible to see from this where the actual load rating limits lie.

New tests on XLPE cable insulations following several years of thermal-electric stress lead, on the contrary, to the statement that measurable *electrical ageing* of the dielectric in the range of practical field strengths up to almost 30 kV/mm takes place considerably more slowly, if at all, than simulated by $N = 12$ [20, 21, 77, 78]. When dimensioning XLPE extra high-voltage cables according to the method described in Section 3.3.2, exponents between $N = 15$ and $N = 17$ are now therefore used in preference [21, 23, 79, 80]. The withstand field strength over 1 hour likewise required for calculating the wall thickness, the so-called design alternating field strength in equation 3.13, is at least 30 kV/mm according to Figure 4.32; other sources even assume 35 to 40 kV/mm [10, 79].

The impulse voltage strength of modern XLPE high-voltage cables has in the meantime considerably overtaken the comparative values of good paper-insulated cables, as can be seen from Figure 4.33. This contains a comparison in the form of two Weibull distributions of the impulse voltage strengths (conductor field strength) at ambient temperature and negative polarity at the conductor of XLPE high-voltage and extra high-voltage cables from recent production and of 400 kV oil-filled paper cables with graded paper layering, taken from Figure 4.9. The maximum design impulse field strength of XLPE required for wall thickness calculation (Section 3.3.2) at which it is certain that no break-

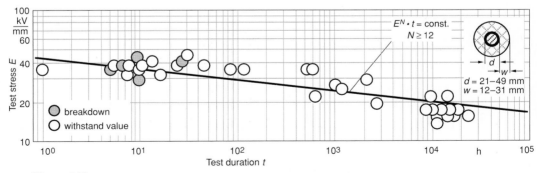

Figure 4.32
Stress/time diagram ("life curve") of XLPE high-voltage and extra high-voltage cables

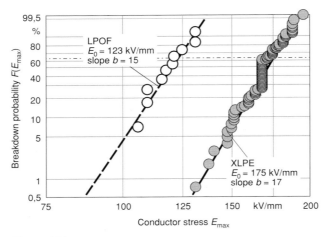

Figure 4.33
Impulse voltage strength of XLPE and LPOF cables at ambient temperature and with negative polarity at conductor

down occurs amounts to 125 kV/mm. The related mean value of voltage/wall thickness (U/w), which is not shown here, is 80 kV/mm [23, 77].

In spite of the superiority demonstrated in Figure 4.33, XLPE cables continued to be dimensioned with somewhat more caution and are in consequence exposed to lower loading in operation than low-pressure oil-filled cables of the same rated voltage. Apart from the life characteristics, which have not yet been conclusively clarified, the reasons for this restraint in utilizing the synthetic dielectric to full capacity are to be found in the fact that the electrical properties are influenced relatively strongly by the temperature and – in the case of unipolar forms of stress such as impulse voltage – also by the polarity of consecutive voltage pulses [81].

As an example of this, Figure 4.34 shows three Weibull evaluations of corresponding impulse voltage tests on XLPE high-voltage cables from earlier trial production. The nominal breakdown strength value of 110 kV/mm achieved at ambient temperature with exclusively negative polarity falls to 90 kV/mm when pulses of alternating polarity are applied and to 70 kV/mm when the conductor is heated to 100 to 105 °C (unipolar loading), thus by more than 35 % compared with the result at ambient temperature. The reason for this behavior is the pronounced capacity of XLPE for storing a charge (Section 3.1.1, Figure 3.3) and the progressive restructuring of the material from a semi-crystalline state to an almost completely amorphous state on approaching the crystallite melting temperature of 110 to 115 °C (Section 4.2.1, Figure 4.23).

Table 4.5 contains a summary of the principal electrical and dielectric (already discussed elsewhere – see Section 3.4.2) properties of XLPE high-voltage and extra high-voltage cables. Attention is drawn to two particular features with reference to the values for dielectric strength listed in this:

- The values of strengths and design field strengths (Section 3.3.2, Determining the wall thickness) specified for identical voltage forms and temperatures vary considerably as, depending on the statistics used, the strength corresponds to a failure probability of

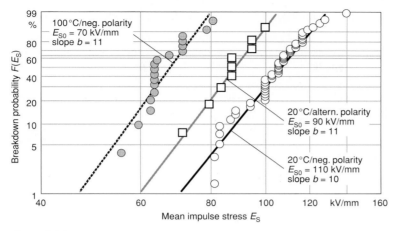

Figure 4.34
Impulse voltage strength of XLPE cables under various test conditions

50...63%, while the design field strengths, by definition, represent withstand values, the failure probability of which is therefore markedly below 1%.

- The value specified for voltage endurance at 50 Hz (>15 kV/mm) is the result of extrapolation of the imaginary life curve in Figure 4.32 to a period of 40 years ($3.5 \cdot 10^5$ h). However, no real results relating to these periods exist as yet.

Table 4.5
Electrical and dielectric properties of XLPE-insulated high-voltage and extra high-voltage cables (extract)

Property	Test conditions	Value
Dielectric dissipation factor $\tan\delta$	50 Hz/20 °C	$0.3\ldots0.5\cdot10^{-3}$
Permittivity ε_r	50 Hz/20 °C	$2.3\ldots2.4$
Dielectric loss coefficient $\varepsilon_r\cdot\tan\delta$	50 Hz/20 °C	$0.7\ldots1.2\cdot10^{-3}$
Specific insulation resistance ρ	20 °C/1 min	$>10^{17}$ Ωcm
Specific insulation resistance ρ	90 °C/1 min	$10^{14}\ldots10^{16}$ Ωcm
Impulse voltage strength at cond. E_{DSmax}	20 °C/1...5/50 µs	$130\ldots200$ kV/mm
Impulse voltage strength at cond. E_{DSmax}	90 °C/1...5/50 µs	$100\ldots150$ kV/mm
Average short-term dielectric strength E_D	50 Hz/20 °C	$45\ldots70$ kV/mm
Average long-term strength E_H	50 Hz/20 °C/40 a	>15 kV/mm
Design impulse voltage field strength at cond. E_{dsmax}	20 °C/1...5/50 µs	$115\ldots125$ kV/mm
Average design impulse voltage field strength E_{ds}	20 °C/1...5/50 µs	$70\ldots80$ kV/mm
Average design alternat. voltage field strength E_{dac}	50 Hz/20 °C/1 h	$30\ldots40$ kV/mm
Average operating field strength E	50...60 Hz	$3\ldots10$ kV/mm
Operating field strength at inner cond. E_{max}	50...60 Hz	$4\ldots16$ kV/mm

Influence of material purity, interface homogeneity and manufacturing technology

Load increases such as have been accomplished in the recent past during the gradual expansion of synthetic and, in particular, XLPE cable technology to cover ever higher rated voltages (see Figure 4.31), would have been inconceivable without a steady improvement in the purity of the material and homogeneity of the interfaces between insulation and semi-conductive layers. Japanese manufacturers thus classify the insulating compounds for XLPE high-voltage and extra high-voltage cables into three *classes of purity* for different rated voltage ranges, with life exponents N of varying magnitudes being assigned to each class on the basis of long-term tests on thin-walled model cables [23]:

- *Super Clean Compounds* for the "lower" high-voltage range up to 90 kV; with life exponent $N = 9$,

- *Extra Clean Compounds* for cables up to 275 kV with $N = 12$ and

- *Ultra Clean Compounds* for the highest level up to 500 kV with $N = 15$ to 17.

Semi-conductive compounds are similarly classified in respect of the homogeneity of their bond with the insulation, with only so-called *Super Smooth Compounds* generally being considered for modern high-voltage cables [74]. The influence of the semi-conductive layer was first investigated primarily on XLPE medium-voltage cables, after certain connections with the quality of the semi-conductive layer had been recognized when clarifying the phenomenon of *water treeing* in particular [72, 82, 83]. Even without

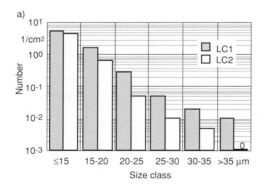

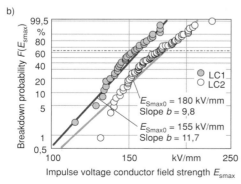

Figure 4.35 Relationship between impulse voltage strength of XLPE cables and quality of semi-conductive layer (according to [72])
a) Density distribution of protrusions at inner semi-conductive layer
b) Weibull distributions of impulse voltage strength

moisture, however, it proved possible to demonstrate direct effects of the interface homogeneity between insulation and semi-conductive layer on important electric properties of XLPE cables.

Reference is made to Figure 4.35 as proof of this statement. Here, the impulse voltage strength of cables with identical insulating compounds but different inner semi-conductive layers LC1 and LC2 is compared with the density distribution of protrusions at the interface between the semi-conductive layer and the insulation as shown in Figure 3.2 (the abscissas description "Size class" here denotes the maximum height of the protrusions of the semi-conductive layer). It is apparent that the smaller number of protrusions per surface unit when using the semi-conductor compound LC2 leads to a significant increase in impulse voltage strength of around 15%, both in the rated value and the minimum value, which is the more important in practice.

In most cases, the effects of increased purity of the insulating material and improved interface homogeneity respectively cannot easily be separated, since advances in technology with regard to the handling and treatment of material during cable manufacture have generally had a positive influence on both features. Thus, by consistently developing each individual process in the manufacture of XLPE extra high-voltage cables, Siemens has, for example, succeeded in just a few years in improving the characteristics which are critical to wall thickness dimensioning to such an extent that cables which were previously rated for 400 kV can now be used for 500 kV without any restriction of their operating reliability [21, 77].

Success is evident here, in particular, in an averaging-out of the product quality to such an extent that the probability of *early failures* is reduced. As an example, Figure 4.36 gives a comparison of two Weibull distributions of impulse breakdown strength (average and maximum field strengths) of XLPE high-voltage and extra high-voltage cables from ongoing production in the periods 1988 to 1991 and 1991 to 1994 respectively. While the respective nominal values do not vary significantly at a difference of 3%, the minimum strength values of the more recent cable specimens surpass those of the older production period in both cases by around 15%. In consequence, it has been possible to raise the design impulse field strengths E_{ds} and E_{dsmax}, which enter into wall thickness dimensioning according to equation 3.15 and 3.16 (Section 3.3.2), from 70 to 80 kV/mm and from 115 to 125 kV/mm respectively (Table 4.5). The distributions, each with at least 100 individual test samples, guarantee the required statistical reliability.

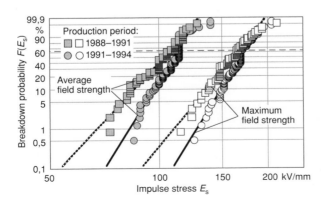

Figure 4.36
Impulse breakdown strength of XLPE high voltage and extra high voltage cable samples from various production periods (Siemens)

As a consequence of the knowledge acquired regarding the influence of material purity, Siemens has in 1996 transferred its entire production of XLPE high-voltage cables to clean-room conditions with graduated standards up to purity class 1000 and at the same time introduced a particularly low-abrasion feed of granular material to the extruders [84, 85] (see also following section).

4.2.3 Methods of manufacture

The manufacture of XLPE high-voltage and extra high-voltage cables basically comprises four fabrication stages, which are listed below in chronological order:

- *Conductor manufacture* (wire drawing, laying up, and braiding if applicable),
- *Core manufacture* (extrusion and cross-linking),
- *Conditioning* (*"vaporization"* of gaseous reaction products in the heat chamber),
- *Sheathing,* with *core screening* taking place beforehand if necessary.

Without doubt, the most important manufacturing phase of these individual stages with regard to the electrical characteristics of the cable is core manufacture, and this consequently forms the focal point of this chapter.

Triple extrusion

It has already been pointed out repeatedly that good electrical stability of thick-walled synthetic insulations in the long term presupposes, in particular, a permanently strong, homogeneous bonding level between the insulation and semi-conductive layers as well as maximum cleanness of the insulating compound used. One milestone along this route is the triple extrusion process first used by Siemens in the 1960s [85], which has now gained acceptance worldwide for cables with a polyolefin dielectric. In this process, the inner semi-conductive layer, insulation and outer semi-conductive layer are extruded in a single operation, using a so-called *triple extrusion head* as shown in Figure 4.37, onto the conductor, which has usually been preheated to around 100 °C, and then jointly cross-linked (see next sub-section). In addition to the firm bond between the three layers of synthetic material, this process in a closed system also offers the advantage that no dust or other foreign particles are able to penetrate the sensitive interfaces with the insulation. *Conductor preheating* reduces the thermal energy to be transferred from outside in the subsequent cross-linking section, as the substantial mass of the conductor metal is itself already at a higher temperature. This facilitates the use of shorter cross-linking sec-

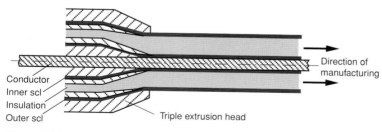

Figure 4.37   Principle of a triple extrusion head

tions and/or higher production speeds in the systems available. As well as this, shock cooling of the molten mass of the inner semi-conductive layer when it comes into contact with the conductor is avoided.

Whereas the semi-conducting compound, which is supplied as a granular material, is normally fed directly into the storage hopper of the related extruder, the insulating compound passes through a number of other stations prior to forming in the extrusion head. The functions of these stations are described in brief below. The details here refer to the processing of granular material which already contains the peroxide and all other additives required for cross-linking on delivery; the *own compounding* practiced by some cable manufacturers calls for some additional components [86]:

- *Material feed and transportation*
  The granulate is delivered by the compound manufacturer in hermetically sealed drums containing a few hundred kg to several tonnes. It must then be fed at a suitable transfer station – the material feed station – into a conveying system, through which it is transported to the extruder. Closed pipelines have proved satisfactory for this in the low-, medium- and "lower" high-voltage range up to around 150 kV. In these, the granulate is transported by a laminar flow of carefully filtered compressed air. The option of discharging the material from an HGV container with a capacity of around 20 tonnes directly into the conveying system is also advantageous, because manual contact with the delivery drum, which always carries a residual risk of contamination, is thus spaced at wide intervals of time.

  Any increase in the distance between the feed station and the extruder increases the likelihood of minute leaks existing in the pipe system and the granular material is also exposed to increased friction on the pipe walls on long transportation paths in spite of a gentle air flow in the pipe. This can consequently lead to the accumulation and gradual ageing of granulate dust at unavoidable discontinuity points, such as joint flanges and elbows. If parts of this dust become detached later, they may get into the insulation. Although these do not signify any hazardous weakening of the dielectric, since they are still insulating, they create certain structural inhomogeneities, which it is important to avoid if possible in the case of maximum electrical loads.

  For these reasons, in 1996 Siemens radically reorganized the feed and transportation of material for XLPE extra high-voltage production in connection with the introduction of clean room conditions [84], Instead of extensive pipe sections laid horizontally, in which the granular material is conveyed by flowing air, the material feed facility, including devices for removing foreign particles, has been installed vertically above the feed hopper of the insulation extruder (Figure 4.38). Not only are optimally short conveying paths thus realized, but any form of mechanically produced air flow is also rendered superfluous: the material flows due to the effect of gravity alone.

- *Air separator*
  The air separator (wind sifter) is a 1 to 2 m long fall pipe, in which the granular material is specifically exposed to a powerful air flow to remove loose, adhering foreign particles, e.g. PE particles removed by abrasion, from the grains of granular material.

- *Magnetic separator*
  The magnetic separator is used to remove particles of magnetic material, such as microscopically small steel tinsel resulting from compound manufacture, for example.

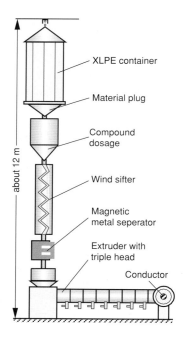

about 12 m

XLPE container

Material plug

Compound
dosage

Wind sifter

Magnetic
metal seperator

Extruder with
triple head

Conductor

Figure 4.38
Vertical arrangement of material conveying and cleaning
system

- *Extruder*
  Here, the granulate is gently melted in several *heating zones* and the molten XLPE
  mass produced is also blended and homogenized to the optimum degree by the rotating *screw*. Precise temperature control inside the extruder is critical to the homogeneity of the insulation: the temperature must be high enough on the one hand to achieve
  a soft, easily formable molten mass, but must not be too high on the other, as then the
  peroxide cross-linking process shown in Figure 4.27 (Section 4.2.1) would commence
  in the extruder prior to forming in the extrusion head. Depending on the molecular
  weight of the compound and the type of peroxide used, the melting temperature on
  entry to the extrusion head should be between around 130 and 135 °C, with even the
  slightest deviations of around 1 to 2 degrees from the recommended value being capable of impairing the product quality. For this reason, an equal number of *cooling zones*
  are also provided along the *extruder cylinder* outside the electric heating devices, to
  enable the automatic control facility to respond quickly to any fluctuations in temperature. Both air and liquid can be used as a cooling medium; in clean room conditions,
  closed systems are always used to avoid turbulence. Figure 4.39 shows the insulation
  extruder for XLPE extra high-voltage cables at the Pirelli Kabel und Systeme cable
  factory in Berlin [84].

- *Strainers in the extruder*
  The molten mass is conveyed to the extrusion head through extruder strainers. A graduated system is used here with varying mesh widths, which become smaller in the direction of flow. The strainers hold back any solid particles in the molten mass, preventing
  them from entering the product as an inhomogeneity.

  The correct design, arrangement and grading of the strainers is often the result of many
  years of process observation and evaluation by the cable manufacturer because, here

Figure 4.39 Extruder with several heating and cooling zones

too, it is important to find the optimum between opposite effects, as with temperature control. The smaller the mesh width, the more reliably the smallest foreign particles – in particular dangerous metal inclusions – are held back. However, the strainer system represents a disturbance (inhomogeneity) in the flow path in principle for the molten XLPE mass, which is semi-liquid due to temperature limiting in order to avoid premature cross-linking. As a consequence, as indicated in Figure 4.40, material deposits can accumulate increasingly in the turbulence zone behind the strainers as production con-

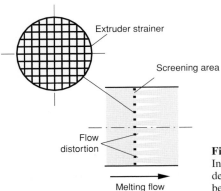

Extruder strainer

Screening area

Flow distortion

Melting flow

Figure 4.40
Inhomogenities in the melt flow through the extruder sieve showing the danger of material deposition behind the narrow points (principle)

133

a) Size distribution

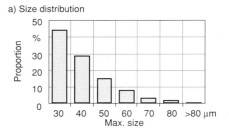

b) Density curve

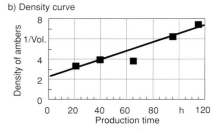

Figure 4.41
Ambers in XLPE cable insulation
a) Size distribution
b) Change in density over production time
(in arbitrary units)

tinues and can age (oxidize) under the long-term influence of the melt temperature of $\geq 130\,°C$ or even start to cross-link. If parts of these deposits then become detached once more in the course of time, they pass unhindered into the insulating sleeve of the cable and there cause structural inhomogeneities in the form of *scorching* or more or less strongly oxidized *ambers*. Although less dangerous than conductive inclusions, these nevertheless still constitute undesirable disruptions of the microstructure in the extra high-voltage range.

Studies with regard to this have in fact shown that the density of ambers (number per unit of volume) increases steadily in XLPE dielectrics as production continues over time, as demonstrated by Figure 4.41 [81]. This shows a typical size distribution of ambers in XLPE high-voltage insulations (Figure 4.41a) and the change in their density over time (Figure 4.41b). As a result, production has to be interrupted at certain intervals to remove the strainers. The maximum production periods which can be achieved depend on the design of the strainer system, the minimum mesh width and the purity requirements in respect of the cable dielectric which, in the absence of binding standards, are the responsibility of the manufacturer alone. In the case of the mesh widths of 50 µm which are common in the extra high-voltage range, a change of strainers is advisable for the most part after around 200 to 250 h, a value which has only been achieved in recent times due to consistent improvements in the area of the extruder and in particular with regard to the flow paths inside the triple extrusion heads.

Dry curing and cooling

In contrast to the manufacture of paper-insulated cables described in Section 4.1.4, the production of XLPE cores has to take place in a single operation without interruption, as the slightest stoppage leads to a build-up of molten mass in the extruder and extrusion head including a dangerous onset of cross-linking. Curing therefore follows on directly from the extrusion process; we therefore refer to continuous curing or *continuous vulcanization, CV.*

As shown in Figure 4.27, the curing process itself starts due to heating of the PE molten mass containing peroxide which has been extruded onto the conductor at around 200 °C. For this purpose, the core is taken directly from the extrusion head to an electrically heated tube, from the walls of which the heat has to be transferred to the core surface.

Up to the 1970s, superheated steam was used for this. This only reaches the required temperature above 200 °C according to its steam pressure curve if the pressure is increased at the same time to approx. 15 to 20 bar. This excess pressure, which must be maintained in the curing tube as well as the subsequent cooling section, at the same time prevents the formation of bubbles in the insulation by gaseous reaction products such as methane (Figure 4.27). Since steam possesses outstanding heat transfer properties, *steam curing* described here constitutes an extraordinarily effective and economical production method. However, one unavoidable side-effect of this cross-linking technique is that considerable amounts of moisture – up to 4000 ppm [87] – diffuse into the hot mass of molten synthetic material and are precipitated during subsequent cooling in the form of small, water-filled voids inside the insulation. These so-called *microvoids* with diameters of up to approx. 20 μm can be clearly recognized in thin sections of newly produced cable as *clouding zones* in the dielectric, as clearly shown in Figure 4.42.

Attempts are now made to avoid high moisture concentrations of this kind where possible, not only because an active role in the occurrence of the undesirable water treeing in PE/XLPE cables is attributed to them (e.g. [82, 88]), but also because they noticeably reduce the impulse voltage strength of new cables [89]. So-called *dry curing* has therefore gained acceptance worldwide, especially for the manufacture of XLPE cables for the highest stresses. In this process, the chemical cross-linking process using peroxides is retained and other media or methods are used to transfer heat from the curing tube to the core insulation: inert gas (nitrogen N2) under pressure or heat-resistant liquid (silicone oil) instead of steam or direct heat conduction from the tube wall to the core surface in which the core cross-section fills the entire cross-linking tube (*Long-Land Die LLD*). In dry cured polyethylene of this kind, the water content of just 100 to 200 ppm [90] remains in the area of the solubility limit of the synthetic material, eliminating the fear of droplet-like precipitations.

Regardless of which type of heat transfer is chosen, uniform heating of the entire insulation layer to curing temperature takes considerable time depending on the thickness. To realize the required residence time of the molten mass in the high temperature region at economically justifiable production speeds, the longest possible cross-linking sections are employed. Depending on the space at the cable manufacturer's disposal and the design of the installation as a whole, heating zones of between roughly 15 and 50 m in length are common. These are followed by a cooling section of at least the same length,

Figure 4.42   Clouding zone in steam cured XLPE cables

in which the insulation – still under external pressure – is cooled at a definably slow rate to values close to ambient temperature.

A production line for XLPE high-voltage cables thus attains lengths of 50 m or more in the area of the heating and cooling zones alone, the core insulation which has been conveyed into these only becoming sufficiently geometrically stable and abrasion-resistant in the final stages following cooling to markedly below the crystallite melting point. The conception and control of the installation must therefore be designed such that deformation of the insulating sleeve and damage to the core surface due to premature contact with the tube walls ("*touch down*") while still soft are avoided completely. The three CV installation types illustrated in Figure 4.43 can be used here, each having specific advantages and disadvantages:

- *Catenary line* (Catenary Continuous Vulcanisation – *CCV system*), the oldest and hitherto most widespread design;

- Vertical line in a *tower* (Vertical Continuous Vulcanisation – *VCV system*);

- *Horizontal line,* developed by the Japanese companies Mitsubishi and Dainichi-Nippon, and thus generally referred to below as the *MDCV system*.

With the exception of the MDCV system, the curing and cooling tubes always have an inner width which is considerably larger than the diameter of the largest XLPE core to be manufactured in the system, so that no touch down occurs when using suitable control technology.

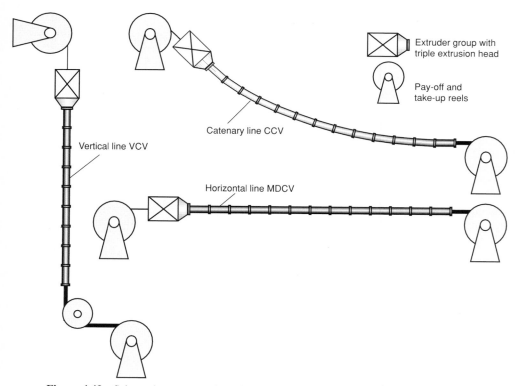

Figure 4.43 Schematic representation of various CV systems for XLPE cable manufacture

Vertical guidance of the core in tower systems offers the most favorable conditions in theory for central positioning of the conductor and exact roundness of the insulating layer, as no unilaterally acting gravitational forces are present. One disadvantage of this type of system is constituted by the often restrictive building regulations, which do not for the most part permit towers of more than 100 m in height, which height is desirable for an economical throughput of material. The operator must therefore be content with shorter curing and cooling sections and take account of this fact by reducing production speeds. Some cable manufacturers have got round this disadvantage in recent times by extending their VCV systems – in some cases even retrospectively – into the ground. However, plant operation remains relatively expensive even then, as all materials still have to be transported first to the highest platform of the tower to process them from there. Older installations also come up against the limits of their loading capacity from time to time when processing larger conductor cross-sections and insulation thicknesses, as the overall weight of the core floating freely in the curing and cooling section has to be supported at the head of the installation.

Horizontal systems circumvent problems of this kind, as here the weight of the cores carried directly in the curing tube (Long-Land Die) can be supported at any time as required. Even the roundness of the core is guaranteed without problem, as the contour is predetermined by the curing tube. Chafe marks on the core surface due to friction with the LLD internal wall are avoided by means of a temperature-resistant synthetic liquid *lubricant* that is pumped from a separate reservoir into the curing tube and creates an effective lubricating film between the core surface and the tube. Horizontal systems also boast advantages owing to the relative ease with which they can be adapted to existing built volumes. Apart from the preferred material feed above the insulation extruder, if applicable, as shown in Figure 4.38, high structures are not required at all, and the length of the curing or cooling sections can be adapted to the production range envisaged on level ground without too great an outlay on construction. The MDCV system was therefore the first choice for numerous cable manufacturers in Europe and the Far East when the need arose in the 1980s to supplement existing production facilities for XLPE-insulated medium and high-voltage cables with a production line for extra-high-voltages (≥ 220 to 500 kV) and large conductor cross-sections (2500 to 3000 mm^2)

Problems of geometry can arise with horizontal systems due to the fact that the copper conductor with its high density of 8.9 g/cm^3 (Table 2.2) sinks into the XLPE molten mass of the insulating sleeve, which is initially still soft (density roughly 0.9 g/cm^3) and thus causes an intolerable eccentricity of the core as shown in Figure 4.44. This phenomenon is countered by three different measures, which are applied individually or jointly depending on the conductor cross-section and insulation thickness:

- Use of insulating compounds with an increased melting index, which form a molten mass that is more viscous and thus more able to bear load

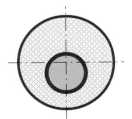

Figure 4.44
Eccentricity of an XLPE core: eccentric position of conductor in round insulating sleeve

- Use of *eccentric tools* in the extrusion head that cause the area below the conductor to be specifically thicker, thus compensating for the effect of sinking

- *Twisting* of the core, i.e. slow rotation around its own axis, so that displacement of the conductor due to gravity is not unidirectional.

An economic disadvantage of horizontal installations is the need keep a wide range of long land dies of high-grade special steel in stock, since a suitable curing tube is required for each core diameter. As well as this, MDCV technology is not optimized for the manufacture of XLPE cores with small diameters owing to the limited length of the curing zone. Manufacturers who wish to cover the medium-voltage range as well as the high- and extra high-voltage level with XLPE cables which are competitive in terms of cost therefore require at least one other type of production installation in addition to a horizontal line.

For the most part, classical CCV-systems are used here. The heating zone and first part of the cooling section following it are shaped as a catenary, a shape automatically taken up by any cable held at both ends and thus also by the XLPE cores on leaving the extruder discharge. Assuming that the automated control facilities are satisfactory, damage to the core surface as a result of premature wall contact can be avoided as safely with catenary lines as in the case of VCV and MDCV systems. In structural terms they are considerably more favorable than tower systems, as the difference in height between the extruder location and take-up for the finished core is only around 15 m; very long heating and cooling sections can be implemented without a large outlay, facilitating high production speeds for smaller core cross-sections in particular. Catenary lines thus represent the optimum production equipment for a broad XLPE spectrum ranging from medium voltage to high-voltage cables with an insulation thickness of roughly 22 to 24 mm.

For wall thicknesses beyond this, core geometry increasingly poses problems, as the still soft molten mass threatens to drop off the freely suspended conductor or – as a preliminary stage to this – assumes an oval contour (Figure 4.45) This *ovality*, mostly in combination with additional eccentricity, proves obstructive for the attachment of prefabricated fittings in particular (Sections 6.2.3 and 7.1) since an exactly round core contour is required for these in most cases. A very effective solution to this problem is provided by Silicone oil continuous catenary vulcanization (SiCCV system), which uses Silicone oil as the medium for heating and covling within the CCV tube. This oil, which has a density practically equal to that of PE, compensates the gravitational effects which cause ovality and eccentricity of the insulation. The techniques used for horizontal systems can also help: the use of eccentric tools and high-viscosity insulating compounds as well as twisting of the core. For this, a process has now been developed further and patented which rotates the conductor prior to it entering the extruder and the finished core when it leaves the cooling section [91]. According to manufacturers' data, it is now possible by

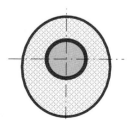

Figure 4.45
Ovality and eccentricity due to downward displacement of molten XLPE mass

means of this so-called *double rotation technique* to use catenary lines for the manufacture of XLPE cores with a wall thickness up to roughly 30 mm. Improvements in the design of the capstans, which have to hold the total weight of the freely sagging core, have also enabled very wide limits to be imposed on the conductor cross-section, so that modern catenary lines are expected to be able to cover virtually the entire product spectrum up to 500 kV.

Apart from the components covered so far – the extruder together with material feed, curing and cooling sections and capstans for transportation of the conductor and core – a complete production line for XLPE cores also comprises the so-called *pay-off* and *take-up*, i.e. reels from which the conductor is paid out or onto which the finished core is wound. Added to these are the conductor preheating by induction already mentioned, a so-called conductor accumulator and various measurement and automatic control devices. The conductor accumulator has the job of being able to "join on" (weld on) a new conductor when the pay-off is empty without interrupting production, although it is only suitable for processing smaller cross-sections of up to approx. 630 mm². The diagram in Figure 4.46 summarizes the most important elements of XLPE core manufacture.

In contrast to the production of paper-insulated cable (Section 4.1.4), the *production speed* with the system configuration specified is determined specifically by the insulation volume and thus by the wall thickness and, to a lesser extent, by the conductor cross-section. The energy which has to be supplied to the molten XLPE mass for the cross-linking process is critical for this (see Figure 4.27). Since only a limited section – the length of the heating zone – is available for this heating process and also the tube temperature cannot be arbitrarily increased, so as to avoid thermal aging of the synthetic material, large-volume cores have to reside for longer in the curing tube or pass through it at a slower speed. Recommended values for modern horizontal systems are 0.8 m/min for 110 kV cores ($w = 18$ mm) with a Cu conductor cross-section of 630 mm² compared with 0.3 m/min for 500 kV cores ($w = 30$ mm) with a 1600 mm² Cu conductor.

The manufacturing process itself only reaches a stable state with a satisfactory product quality after a start-up phase of varying length. The so-called *entry length*, in particular, which passes through the initially unpressurized cross-linking and cooling tube is wasted (reason: the external pressure required to avoid bubble formation in the dielectric can only be built up if the conductor is surrounded over its entire length by insulation, as the pressure-generating medium otherwise escapes through the multistrand conductor). The same applies to the end of production: here too the excess pressure has to be switched off before

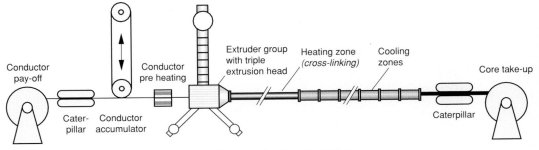

Figure 4.46   Principal components of a production line for XLPE cores

turning off the extruder, so that once again a full system length is lost (*exit length*). Economical production therefore depends on the maximum machine operating time between the feed and exit length.

Sheathing

In contrast to the manufacture of paper-insulated cable, in which sheathing directly follows core manufacture to avoid air and moisture being reabsorbed by the dry, impregnated insulation (Section 4.1.4), XLPE cores must first be subjected to a fairly long conditioning phase. In the course of this, some of the reaction products arising during curing, in particular flammable methane gas (Figure 4.27) are *vaporized* out of the insulation sleeve and the multi-strand conductor by the application of moderately increased temperatures (approx. 60 °C). This measure is intended primarily to prevent incalculable deflagration during the subsequent installation of fittings on site, e.g. due to welding or soldering. The duration of the conditioning process is oriented to the core geometry and is between one and four weeks for high-voltage and extra high-voltage cores.

The subsequent stages in completion of the core to form a finished cable are not noticeably different to the operations described in 4.1.4 for paper cables, provided that the XLPE cable has a metal sheath (lead or aluminum) with corrosion protection. Equation 4.8 ($D = d + 2s + 2t$) for determining the outer diameter of a corrugated aluminum sheath should be recalled here, as particular importance is attached to the expansion gap s taken into account in this equation for avoiding mechanical damage to the core surface due to heating for the XLPE with its extremely large thermal expansion coefficient. To give extra protection, a permanently elastic, conductive padding material is spun around the core prior to production of the corrugated sheath, this material ensuring uniform ohmic contact between the outer semi-conductive layer and the sheath. Figure 4.47 shows an XLPE corrugated sheath cable for 400 kV as an example.

In Japan, stainless steel has also been used widely instead of aluminum as a sheath material for XLPE extra high-voltage cables in a similar construction. Low weight and a more favorable reduction factor are cited as advantages, although these are balanced by markedly higher costs compared with aluminum.

One special feature which is only feasible for solid cables is offered by synthetic sheaths in combination with aluminum foil glued axially in the longitudinal intake as a vapor barrier or *lateral water protection* as shown in Figure 2.7 (Section 2.3.4). These laminated sheaths now generally consist of (thermoplastic) high-density polyethylene (HDPE), which is distinguished chiefly by its high mechanical abrasion resistance and which assumes the function of corrosion protection for the aluminum foil at the same

Figure 4.47
400 kV XLPE cable with corrugated aluminum sheath

time. Owing to the small cross-section of the metal foil, its current carrying capacity is negligible, so that laminated sheath cables have to have a copper wire screen in principle, which to avoid *screen wire impressions* in the core surface on heating, is applied over so-called *screen bedding* in the form of conductive tapes or non-woven fabric.

Although the presence of the metallic lateral water protection reliably keeps moisture out of the insulation if the sheath is intact, mechanical damage to the insulation due to external influences can never be entirely ruled out. To ensure that possibly larger cable sections are not damaged in the long term as a result by water penetrating from outside, additional *longitudinal water protection* is always provided on laminated sheath cables, and in many cases even where a solid metal sheath is provided, in the screen area or between the core and metal sheath. Normally this takes the form of expanding tapes or non-woven fabric, which swell due to the effect of moisture and seal the interstices between the individual screen wires or other gaps between the core and the sheath. The possible consequences of a sheath leak are thus limited to the immediate vicinity of the damaged area. In the case of corrugated aluminum sheaths, longitudinal water tightness can only be achieved in conjunction with annular corrugation, not with screw corrugation.

The screen bedding, wire screen and expanding non-woven fabric or tapes are applied following vaporization of the XLPE core in principle in a single operation before XLPE cable fabrication is completed in a further common stage by the longitudinal intake of the polymer-coated, self-fusing aluminum foil and the polyethylene sheath extrusion. Production lines also exist which apply all the structural elements present over the core, ranging from the screen bedding to the outer sheath, in a single manufacturing step. In this case, however, so-called *SZ stranding*, which poses more of a problem in terms of the uniformity of the wire intervals, has to be used instead of conventional circulatory screen wire winding to avoid production stoppages while the winding machines are refilled [92].

Laminated sheaths primarily cover the XLPE cable segment up to 150 kV, but can also be used in principle for higher rated voltages where the conductor cross-sections or weights are not too great. Lead or corrugated aluminum sheaths are preferred above 150 kV for mechanical reasons and to withstand large short-circuit currents. Further features of the different sheath types are outlined in Section 2.3.4, Table 2.4. Figure 4.48 shows a specimen of a 110 kV XLPE cable with a laminated sheath.

Figure 4.48
110 kV XLPE cable with laminated sheath

5 Bulk-Power Transmission Systems

It was established in a variety of different ways in Section 3.5 that the transmission capacity of conventional cable systems with natural cooling falls significantly short of that of overhead lines of the same nominal voltage, in particular because of the limited heat dissipation through the soil (see Figures 3.37 and 3.38). This means that in order to transmit electrical energy that has been brought in through an overhead line, several parallel cable systems are required. It is possible to achieve something approaching the load capacity of overhead lines (or even a reversal of the capacity ratio) by using what are known as *bulk-power cables*. This term is used to describe cable systems whose heat losses are either removed by artificial cooling or are drastically reduced by the use of alternative techniques. The following are specific examples of techniques included in the latter category:

- Conventional high voltage cables with *forced cooling* (Section 5.1),

- *Gas-insulated lines GIL* ("SF$_6$ cables", Section 5.2) and

- *Low temperature cables* based on *cryocables* with normal conductor or *superconductor cables* (Section 5.3).

These three alternatives are described in succession in the main section above, followed by a comparison that takes into account technical, economic and ecological angles.

5.1 Force-Cooled Conventional High Voltage Cables

5.1.1 Fields of Application for Force-Cooled Cables

It was pointed out in Section 3.5, with the help of a number of examples (e.g. Figures 3.37 and 3.38), that conventional, naturally-cooled cable systems can at best achieve barely 50% of the transmission capacity of overhead lines of the same voltage level. If, therefore, the electrical energy delivered by overhead lines is to be taken further in conventional cables and with the same transmission performance, then the only remaining option is the installation of double or multiple parallel systems and/or the use of artificial cooling.

Particularly within areas of dense population, the preferred area of application of cables, it is often impossible from the outset to consider installing multiple parallel high voltage cable systems since the required widths of cable run are not available. But even in situations where the spatial conditions do permit this type of solution, from the point of view of technical implementation it still remains questionable. Since the transmission capacity of a cable system is limited primarily by the finite heat dissipation through the ground (as explained in Sections 3.4.2 and 3.5.3 – Formula 3.32), the result is mutual thermal interference when several systems are bundled together in one route: The combined heat flow resulting from the power losses from all the cables – even when trenches are back-filled with thermally stable material – results in widespread drying and build up of the thermal resistance of the surrounding soil. For this reason, the losses that can be dissipated and consequently also the maximum transmission capacity per cable system as calculated in Formula 3.32 fall far short of the values for a single system laid in the same environment. The power gain achieved by using several parallel sections of cable is thus limited, and bears no relation to the increased expenditure.

The use of artificial cooling, on the other hand, has the effect of significantly increasing the losses that can be dissipated, and consequently also the transmission capacity of a cable system. As demonstrated in the next chapter, increases of 100 to 200% can be achieved without unreasonable expense. There are also the additional advantages of forced cooling in that, when the appropriate method is used, the problem of heat accumulation caused by several parallel cable systems does not arise, and that the cooling performance can be easily adjusted to suit the current load on the cable. High-capacity sections for the supply of densely-populated areas, often forming a continuation of an overhead line, thus form the preferred application field for artificially-cooled conventional high voltage and extra high voltage cables.

5.1.2 Cooling Methods

The role of forced cooling is to dissipate the lost power that has been converted to heat in the structural metal components of the cables – i.e. most importantly in the conductor – as close to its source as possible, and thus to hold the conductor temperature below the insulation-dependent maximum levels (see Table 3.2) regardless of the composition and moisture content of the surrounding soil. Assuming that all the lost heat passes into the cooling medium and is carried away by it, then the maximum permissible loss capacity P'_{ab} of the cable system equates to the following per unit of length [67]:

$$P'_{ab} = (c_k \cdot Q_k \cdot \Delta \vartheta) / l_k \tag{5.1}$$

where

c_k = Specific thermal capacity of the cooling medium in Ws/K
Q_k = Coolant flow (mass rate of flow) in m^3/s
l_k = Length of cooled section in m
$\Delta \vartheta$ = Difference between exit and entry temperatures of coolant in K.

It follows from the c_k values of the most suitable coolants in Table 5.1 that gases – because of their low thermal capacity with realistic temperature differences of some tens of K – require either extremely high-volume flow or are suitable only for very short cooling sections. With the exception of special applications, e.g. the forced-air ventilation of cable tunnels or the cooling of "hot spots" in gas pressure cables, liquid media will therefore continue to be relied on for the dissipation of heat.

Table 5.1 Specific thermal capacity c_k of certain coolants

Medium	c_k in Ws/K
Air (1 bar)	1.3
Nitrogen (15 bar)	12.5
Cable oil	1750
Water	4180

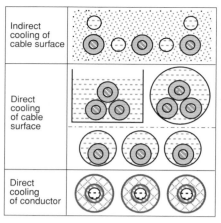

Indirect cooling of cable surface		Conduits buried in parallel to the cable system; cooling agent: water
Direct cooling of cable surface		3 cables in open trench natural slant with dams; cooling agent: water. 3 cables in one pipe forced circulation; cooling agent: water
		Cables in separate pipes forced circulation; cooling agent: water
Direct cooling of conductor		Hollow conductor with enlarged diameter forced circulation; cooling agent: water or oil

Figure 5.1

Methods of forced cooling in high voltage cables (as described in [67])

A distinction is made between three fundamentally different cooling methods according to the point of application in the cable and the means of heat removal; these methods can be further differentiated (see Figure 5.1), e.g. by the use of different thermal transfer media or by changing the way in which coolant is circulated:

- Indirect surface cooling (or lateral cooling)

- Direct surface cooling (integral cooling)

- Direct conductor cooling (internal cooling)

Indirect cooling of the cable surface

As shown in Figure 5.1, the cable surface is cooled indirectly through the installation of pipes carrying water that are laid parallel to the cable system and in the same trench. These pipes are generally made from heat-resistant HDPE, although in principle they can also be manufactured in non-magnetic material. In order to reduce the heat transfer resistance between cables and cooling pipes they are laid together in lean concrete or other bedding material with a similarly useful ability to conduct heat. The lost heat absorbed by the water in a particular section of cable (the *cooling section*) is removed in cooling stations by the use of heat exchangers.

Greatest efficiency is achieved when the water flows first out of the heat exchanger and through the two pipes lying between the phases in a *closed circuit*, and is then routed back to the cooling station through the two pipes running above the outside cable (see Figure 5.2). Most important of all, this *counterflow principle* enables a more even distribution of temperature to be achieved along the cable section than would be the case with only unidirectional flow (*open circuit*) [93]. A critical factor determining the effectiveness of forced cooling is the temperature increase $\Delta \vartheta$ along the cooling section as defined in Formula 5.1, i.e. the initial and final water temperatures ϑ_e and ϑ_a.

The advantage of this type of forced cooling lies in the straightforward nature of the technology. The laying of the cables and the installation of the accessories are exactly the same as the corresponding operations for naturally cooled systems. The HDPE cooling pipes – like the cables themselves – can be wound on transportation reels and thus deliv-

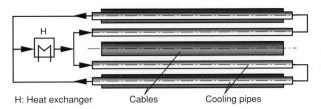

H: Heat exchanger　　　　Cables　　　　Cooling pipes

Figure 5.2　Counterflow principle in indirect surface cooling with four cooling pipes

ered to the site in longer sections; the procedure for laying them is virtually the same as the laying of cable. Another excellent characteristic of indirect surface cooling is its great operational flexibility. Thus, not only can the cooling performance be easily tailored to suit the requirements concerned, but it is also possible without great expense to set up systems in such a way that they can "grow". Cables and accessories are installed in the normal manner, and the cooling pipes are laid parallel to them, but without initially setting up the pump stations and heat exchangers required for the cooling operation. Only if, at a later point in time, the actual loading on the cable necessitates artificial cooling will the missing components be added and commissioned. This approach was used, for example, with the 400 kV oil-filled cable system installed by Vienna municipal services [94].

These advantages are offset by the limited effectiveness of indirect surface cooling, which can be attributed primarily to the finite thermal resistance that still exists between the cable surface and the coolant. This can lead to the occurrence of local high temperatures ("hot spots"), particularly at crossover points with other cable runs or district heating pipes. 'Hot spots' adversely affect the current carrying capacity of the entire system, and may require supplementary cooling by *heat pipes*, as explained in Section 3.5.2.

When indirect surface cooling is employed, the transmission capacity of a cable system can be increased by around 50 % to 120 % [43, 95] relative to the conditions with natural cooling, as shown in Fig 3.37 (Section 3.5.3). The exact transmission capacity depends on the design of the system and the temperature difference $\Delta\vartheta = \vartheta_a - \vartheta_e$ within the separate cooling sections as per Figure 5.2 and Formula 5.1. By way of example, Figure 5.3 shows a comparison of the transmission capacities of oil-filled cable runs for 110 and 400 kV.

Direct cooling of the cable surface

This principle involves bringing the cable surface into direct contact with the coolant. The thermal resistance between cable surface and coolant, which plays such an important role with indirect cooling, is thus not an issue here and as a result this method becomes more effective. Figure 5.1 contains sketches of three possible versions of direct surface cooling:

• All three phases laid together in an open water-filled trench with a natural slope,

• All three phases laid together in an enclosed pipe with coolant flowing through it. Here, depending on the cable type, other thermal transmission media apart from water might be considered, such as oil with high-pressure oil-filled cables or nitrogen with gas pressure cables, and lastly

145

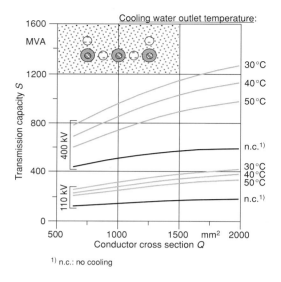

Figure 5.3
The effect of indirect surface cooling on the transmission capacity of oil-filled cable systems for 110 kV and 400 kV

- The three phases laid separately, each in its own cooling pipe, in which case water is generally reverted to as the coolant because of its convenient thermal capacity (Table 5.1) and because of its harmlessness from the ecological point of view. In terms of technology as it stands at present, this third method represents the most effective of the direct surface cooled transmission systems. However, unlike the solution in which the phases are laid together in a pipe, non-metallic coolant pipes do have to be used for this solution in order to avoid additional induction losses. To achieve optimum heat removal, the cables should be arranged in a flat configuration and the incoming cooling water from the heat exchanger should first flow simultaneously through the two cooling pipes of the phases on the outside before passing back through the central pipe to the heat exchanger at double the flow rate (see Figure 5.4). In a similar manner to the case of indirect cooling, here, too, this counterflow principle has the effect of producing a uniform cable temperature along the cooling section.

The most significant advantage of direct over indirect surface cooling is the lack of thermal transfer resistances between cable surface and coolant. As long as the coolant is able to remove sufficient heat, the limit to the transmission capacity is determined purely by the thermal resistances between the conductors and the surface of the jacket within the

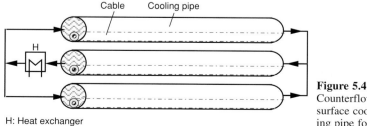

Figure 5.4
Counterflow principle in direct surface cooling with one cooling pipe for each phase

H: Heat exchanger

146

Table 5.3
Technical specifications of implemented extra high voltage cable systems with forced cooling

System	Bewag Berlin I (Germany)	Bewag Berlin II (Germany)	Vienna (Austria)	Dinorwic (GB)
Commissioning, source	1978 [96]	1994 [97, 105]	1979[1] [94]	1979 [93]
Cable, nominal voltage	400 kV LPOF cable with Al corrugated sheath			
Phase configuration	flat formation, cross bonding		flat formation in lean concrete, cross bonding	
Length in system-km	2 x 8	2 x 8	2 x 12[1]	1 x 11, 1 x 6
Max. capacity in MVA/system	1100	1100	1100 (600 without cooling)	1770
Conductor cross section in mm²	1200 Cu	1200 Cu	1200 Cu	2500 Cu Milliken-Cond.
Surface cooling	direct water	direct water	indirect water	indirect water
Cooling pipes/system	3 asb. cement	3 HDPE	4 HDPE	4 HDPE
Cooling circuit	Counterflow enclosed			
Joint cooling	Oil circulation in hollow conductor		additional surface cooling	

[1] Extended in 1984 by 2 x 6 km, and in 1986 by 2 x 4 km, using identical technology

Figure 5.8
Sample of one phase of the 400 kV oil-filled cable in the Bewag II system with direct surface cooling in the HDPE pipe [97]

Figure 5.8 shows the arrangement of the 400 kV low-pressure oil-filled cable used for the Bewag II system in the HDPE pipe used for direct surface cooling. This type of pipe had to be installed for environmental reasons in place of the asbestos-cement pipes used in the earlier Bewag I system [105]. Figure 5.9 gives an idea of the layout of cables and cooling pipes in the trench for the indirectly-cooled 400 kV system installed by Vienna municipal services.

Figure 5.9
Arrangement of cables
and cooling pipes in a
section of 400 kV oil-
filled cable system
installed by the Vienna
municipal services [94]

In the highest capacity range with transmission currents in excess of 2000 A, e.g. with capacities in excess of 1500 MVA at 400 kV, alternatives to conventional cables with forced cooling are offered by the various techniques of underground energy transmission listed earlier at the beginning of Section 5:

- *G*as-*I*nsulated *L*ines *GIL* and
- Low-temperature cables

5.2 Gas-Insulated Lines GIL

The development of the modern construction of gas-insulated transmission lines is based on the technology of gas-insulated switchgear. These lines have been used since the 1970s as busbars in high voltage switchgear as connections between switching stations with one another or to link transformers into switching stations. The majority of applications to date have been within switching stations, and thus necessarily over short lengths (Figure 5.10).

Figure 5.10
Surface-installed double section of a 154 kV SF$_6$-insulated tubular conductor system (Furukawa Electric, Japan)

5.2.1 Rigid Tubular Conductors

These pipe systems are either single-phase or three-phase and are enclosed in an outer pipe. Because of the large diameter, very few three-phase enclosed pipe conductors have so far been installed with rated voltages above 300 kV [106, 107].

The single-phase tubular conductors consist of rigid aluminum tubes arranged concentrically, representing the grounded housing (sheath) and the conductor (Figure 5.11). The diameter ratio of tubular conductors is normally 2.5 to 3.5, i.e. close to the theoretical optimum of $R/r = e \approx 2.7$. Thus, the external diameter of single-phase tubular conductors is approx. 3 to 4 times greater than in conventional cables.

The inner conductor is secured using *cast-resin post insulators* or *disc insulators*, the latter serving simultaneously to shield separate gas compartments. Until now, only *sulfur hexafluoride* (SF$_6$) has been used as an insulating gas, and – in accordance with switchgear technology – is maintained at an operating pressure of 3 to 6 bar. For reasons of transportability the separate factory-produced sections of pipe which have to be screwed or welded on site are limited in length to between 10 and 12 m. For longer runs this results in a large number of joints, which in turn means a correspondingly greater risk factor.

The post insulators and disc insulators are the most critical components in a cable system with pipes: They are subjected to a relatively high electrical stress and must be capable of withstanding the operating field strength and all the stresses involved in testing – both in terms of volume and the surface area adjoining the surrounding gas compartment (danger of creepage discharges!).

These demands are met in the case of *disc insulators* by choosing a funnel-shaped design, which, most importantly, encourages a reduction of the dangerous tangential field with the specified cylinder configuration (see Figures 5.11 and 5.12). *Post insulators* usually have a surface featuring cast resin ribs in order to make the creepage path longer (Figure 5.12). Particular attention must be paid to the contact points between the insulators and the inner and outer conductors in order to avoid pre-discharges in unavoidable gaps. The surface of the insulator that is in contact with the inner conductor is therefore always provided with a conductive coating, as a result of which any gap that may be present is moved to the field-free area. Additional stress relief can be achieved by so-called "*prominent electrodes*" at conductor or ground potential; such electrodes reduce the field strength by a specified amount in the vicinity of the clamp points.

As a result of the temperature differences between no-load operation and full load, the pipe elements can change in length; this effect must be counteracted by constructional measures. For short runs the conductor tubes are housed flexibly in *ring contacts* (Figure 5.13), while for sheath tubes *convoluted expansion joints* must be used (Figure 5.14).

According to the manufacturers' specifications (e.g. [108]), standard modules can be combined to implement any conceivable transmission route. This is generally achieved

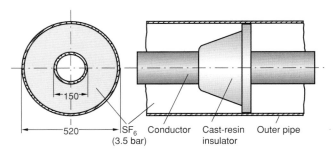

150

520 SF₆ (3.5 bar)   Conductor   Cast-resin insulator   Outer pipe

Figure 5.11
Construction of a SF$_6$-insulated tubular conductor for 420 kV (schematic)

Figure 5.12
Disc insulator and one-sided cast-resin post insulator

through the use of special *ball components*, as can be seen in the foreground of Figure 5.15. These components need a much larger diameter than the adjacent tubular conductors in order to withstand the specified operational and test voltages. In electrical terms they can almost be regarded as concentric spheres, and the field inhomogeneity within them is much greater than in coaxial cylinders.

The first gas-insulated tubular conductor system approx. 100 m in length was constructed in the USA in 1971 for a capacity of 1200 MVA at 345 kV. The first 420 kV system in Europe for 2500 A nominal current was successfully commissioned in Germany in 1975 as an outgoing power line for an underground power plant, and the length of the system installed was 700 m [108]. This system is even now one of the longest in the world (Figure 5.16). Thus, for single-phase enclosed rigid tubular conductor systems there are now more than 20 years of operational experience to use as a basis; systems of this type are generally viewed in a positive light.

In Nagoya (Japan) a GIL system for 275 kV and a transmission capacity of 2850 MVA is currently being installed in a tunnel, and its length of 3.3 km will make it the longest in the world. This single-phase enclosed double system is scheduled for commissioning in 1998.

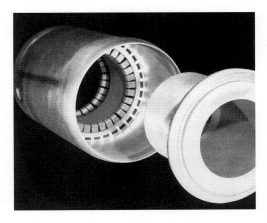

Figure 5.13
Conductor plug-in connection with ring contacts

Figure 5.14
Axial expansion joint for the outer pipe (convoluted expansion joint)

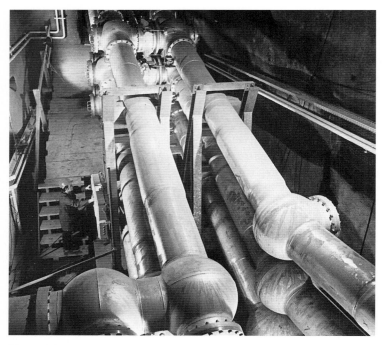

Figure 5.15
420 kV tubular conductor with ball components for changing direction (as described in [108])

Figure 5.16
420 kV GIL as a double-system line from an underground power plant [108]

The state of the art in gas-insulated tubular conductor systems was a keynote topic at Cigré 1984 [109] as well as at the Jicable conference in Paris in 1995 [106]. Amongst other things the following points were reported:

- The number of systems set up before 1993 was around 130 with a combined total system length of approx. 30 km. More than half of these systems are ≤ 200 m long; the average length is 250 m, and the longest 700 m.

- In the majority of cases, gas-insulated lines are laid in air on the surface (Figure 5.10) or in open trenches. Tunnel installations account for around 20%, and are used predominantly for outgoing lines from underground power plants. The proportion of underground tubular conductors is smaller still (< 15%).

- The application range for tubular conductor systems starts at 138 kV and extends to 550 kV, or in individual cases even up to 800 kV. More than 50% of the systems lie in the voltage range above 300 kV.

- The transmission capacity per system ranges from 300 MVA at 138 kV to 7600 MVA at 550 kV. Table 5.4 shows the average and maximum transmission capacities of systems implemented to date. These figures confirm the fact that extremely large amounts of power can be transmitted over short distances using modern gas-insulated tubular conductors. The fact remains, however, that modern tubular conductor technology ought to be viewed as a modified switchgear technology rather than the perfect technical solution for long-distance underground power transmission [106].

- By way of example, Table 5.5. shows typical data for theoretical and actual tubular conductor systems. One striking difference is the large external diameter in comparison with conventional high voltage cables. The aim is to reduce these diameters and consequently also the costs by increasing the operating stress of the cast-resin insulators [110, 111].

Table 5.4
Average and maximum transmission capacities of existing tubular conductor systems laid above ground [106]

Voltage in kV	< 200	≥ 200 … < 300	≥ 300 … < 400	≥ 400
Average transmission capacity in MVA	500	1350	1400	3150
Max. transmission capacity in MVA	1050	3400	2500	7600

Table 5.5 Examples of single-phase enclosed rigid tubular conductor systems [108, 110, 111]

Nominal voltage in kV	154	275	420	550	550
Conductor diameter in mm	100	160	150	230	180
Internal diameter of outer pipe in mm	340	480	530	700	480[1]
SF_6 pressure in bar	3.5	3.5	4	3.5	6
Nominal current in A	2000	4000	2500	8000	8000
Transmission capacity at laying in air in MVA	550	2000	1800	7000	7000

[1] The dimensions are reduced by increasing the operating stress of the cast-resin insulators from 3 to 5 kV/mm

Table 5.6
Comparison between SF_6-insulated tubular conductors and XLPE-insulated extra high voltage cables

Advantages of tubular conductors over XLPE cables	Disadvantages of tubular conductors compared with XLPE cables
• Large cond. cross sections are possible, allowing high transmission powers (especially when laid in air)	• Larger external diameter of pipe, therefore wider route path
	• Only rigid pipes possible at present, therefore limited transport lengths
• Good heat dissipation to the surroundings due to large diameter of pipe	
	• Many connection points needed on site; therefore increased risk of impurities in the gas compartment
• High overload capacity	
• Insulation medium not easily flammable	
	• Compensation is required for the axial thermal expansion of the pipes
• Minimal dielectric losses	
	• Costly system components for changes in direction when the route is not straight
• Low current-dependent losses	
• Lower operational capacitance, therefore lower charging power; no reactive power-compensation required	• Problem of environmental compatibility of SF_6
	• Necessity for gas monitoring
• Simple transition to SF_6-insulated switchgear	• Higher costs

Table 5.6 contains a summary of the advantages and disadvantages of SF_6-insulated tubular conductors in comparison with polymer-insulated (XLPE) extra high voltage cables.

5.2.2 Flexible SF_6-insulated cables

A serious drawback to the tubular conductors described above is their rigid construction. Tests have therefore been performed in Germany [112], England and the USA [113, 114] to explore the possibilities of manufacturing flexible SF_6-insulated tubular systems using the corrugated sheath technology that has been successfully tried and tested in high-frequency cables and long-distance heating lines. The necessary degree of flexibility to allow transportation on reels is achieved through the use of a stranded conductor covered with slightly corrugated copper tubing, along with a corrugated outer sheath [Figure 5.17]. Disc-shaped plastic spacers, positioned on the conductor immediately before the sheathing, secure the conductor concentrically relative to the outer sheath (Figure 5.18).

The first stage in the sheathing process is to form an axially presented metal band made of aluminum alloys, copper or non-magnetic steel into a tube around the conductor and to seal-weld it. The tube then passes into a corrugating device where its surface is corrugated. This procedure can be used to produce tubes up to 450 mm in diameter, which can then be wound onto drums and transported [112]. However, the maximum drum size (limited in Germany, for example, to approx. 4.5 m in diameter by the rail or road profile) results in a limit to the length that can be transported of between approx. 100 m and 200 m.

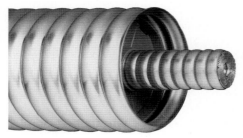

Figure 5.17
Flexible SF$_6$-insulated cable for 220 kV [112]

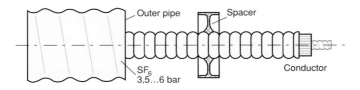

Figure 5.18
Design of a compressed gas insulated cable with a corrugated sheath (schematic representation)

The level of voltage that can be achieved with a flexible SF$_6$ cable depends on the cable diameter and the SF$_6$ pressure; both of these factors have an upper limit. For testing purposes, SF$_6$ cables were produced in 110 kV, 230 kV and 345 kV versions. Table 5.7 contains the typical specifications of these cables. The 345 kV cable was tested at the Waltz Mill cable testing station in the USA [114]. With an external diameter of 390 mm and approx. 4 bar SF$_6$ pressure, it proved possible to meet the requirements specified for the 345 kV level. The extrapolation of these results to 500 kV gives an external diameter of approx. 550 mm with the same SF$_6$ pressure.

The problems around the post insulators or *spacers* – which were emphasized above from the high voltage point of view with regard to rigid tubular conductors (Figure 5.12) – become more evident with this technology because of the increased field strength result-

Table 5.7 Characteristics of flexible SF$_6$-insulated cables [112, 113]

Nominal voltage in kV	110	230	345
Conductor diameter in mm	80	120	120
Corrugated sheath diameter in mm	250	330	390
Transportable length[1] in m	180	100	80
SF$_6$ pressure in bar	3.5	4	4
Nominal current in A	2500	1900	2000
Max. transmission capacity when installed in air in MVA	500	750	1200

[1] With 4.5 m drum diameter

ing from the corrugation of the conductor and outer tubing. Furthermore, problems are encountered in designing and fixing the spacers on the corrugated pipes. There is a danger, e.g. when laying the cable, that movements of the pipe will cause the formation of gas gaps between the spacers and the pipe wall in which partial discharges may occur under operational or test voltage.

Because of these limitations due to physical factors and technical factors associated with high voltages, it proved impossible to develop flexible SF_6-insulated cables as an alternative to rigid tubular conductors. They have never yet been used commercially for bulk-power transmission world-wide.

5.2.3 GIL for Long Transmission Distances

As explained above, international experience with modern tubular conductor technology has shown that it is possible to achieve high-capacity transmission, but only over short distances and at relatively high cost.

GIL technology therefore needs to be improved considerably from its present stage in both technical and economic respects in order to be considered as an underground alternative to overhead lines over long distances, e.g. on the 400 kV level. Experts are of the view that there is potential for such improvement [106, 115, 116]. However, further development requires considerable research and development investment in order to improve environmental compatibility and reliability, and to reduce costs.

Electricité de France (EdF) presented a research and development programme with this aim at the 1995 Jicable conference in Paris; it is being carried out in partnership with manufacturers and aims to establish whether or not it is possible to meet the requirements of the 400 kV level through new GIL technology [106]. The objectives are 400 kV transmission systems capable of transmitting 2000 to 3000 MVA underground over distances of up to 100 km, while costing less than ten times an equivalent overhead line. Because of the high levels of investment involved, there is a service life requirement – as is common for conventional high voltage and extra high voltage cables – of at least 40 years. The following aspects were also addressed:

- In order to improve the environmental compatibility of the GIL, it is anticipated that SF_6 as the insulating gas will be replaced by pure nitrogen [106] or by a mixture of nitrogen and SF_6. For dielectric reasons, the optimum proportion of SF_6 appears to be <20%. If pure nitrogen is used as an insulating gas, the operating pressure must be increased to approx. 10 bar and the diameter of the outer pipe with single-phase enclosure of a 400 kV system (specified withstand lightning impulse voltage $U_{rB} = 1425$ kV) increased to approx. 700 mm.

- The large pipe diameters and the use of gases as a dielectric mean that the GIL systems exhibit a low operational capacitance in the order of magnitude of 50 to 70 nF/km. This results in a critical transmission length of 400 to 500 km, i.e. longer than the standard 400 kV links using overhead lines (see Section 2.2.1, Figure 2.1). In this respect, GIL technology, unlike the conventional 400 kV cables, represents a real alternative to the overhead line.

- Careful attention must be paid to the thermal capability of the GIL with these high transmission capacities with regard to the temperature of the conductor and the outer pipe and also to the mechanical stresses caused by the elongation of the components.

160

In order to avoid the soil drying out when laying pipes directly in it, the surface temperature of the outer pipe – as with conventional cables – must be limited.

- As well as the single-phase enclosed tubular conductors, the three-phase enclosed version is also open to discussion. The main problem with this technology lies in the control of the short-circuit forces in the conductors, which are routed in parallel in the shared outer pipe. On the other hand, a three-phase enclosed tubular conductor does also offer a number of advantages over three single-phase pipes:

  - virtually no reverse current flows in the outer pipe, i.e. the losses are smaller
  - the necessary width of route is less
  - the production costs are lower.

  These factors combine to give lower system costs.

  [107] describes a three-phase enclosed tubular conductor for 420 kV and 2000 MVA transmission capacity. The outer pipe has a diameter of approx. 1.2 m; nitrogen under a pressure of approx. 10 bar is used as the insulating gas.

The three main goals in developing a new GIL technology and the key areas of focus to emerge from the development are described in Table 5.8.

The EdF test program was launched in 1996 in their test laboratory "Les Renardiéres". Prototypes from 10 to 35 m in length have passed the initial electrical and thermal performance tests. Development should be completed by the turn of the century, and it should then be possible to answer the question of whether the new GIL technology can meet the requirements for power transmission systems on the 400 kV level in terms of reliability, environmental compatibility and cost. If the outcome is positive, the plan is to immediately start construction of the first short pilot underground 400 kV system (which will be a few km in length).

At the 1997 IEEE/PES-ETG conference in Berlin, Bewag/Berlin presented a pre-qualification program for tunnel-laid 400 kV GIL systems. The intention is for various differ-

Table 5.8
Development goals and key points for new GIL technology [106, 116]

Development goals	Key points
1. High level of environmental compatibility	• Direct underground installation ("invisible") • Environmentally compatible insulating gas • Little electromagnetic interference
2. Improvement of current technology	• Single-phase and/or three-phase housing • Ability to transmit power over long distances • Reliable electrical dimensioning for other insulating gases • Reduction of current-dependent losses • Mechanical and thermal behavior with large currents and direct underground installation • Simplification of system components for directional changes • Installation on site under clean-room conditions • Tests after installation
3. Reduction of costs	• Optimization of production on site • Implementation of direct underground installation using pipeline technology

ent manufacturers to carry out tests, including an endurance test with 480 kV over 2500 h, each manufacturer using a separate 60-m-long test route. The twin aims of these studies are to test the engineering involved in installing the systems in a tunnel (3 m diameter for two systems) and to prove the reliability of all the components.

5.3 Low-Temperature Cables

Another alternative to force-cooled cables that should be considered for bulk-power transmission is the so-called *low-temperature cable*. The fundamental principle behind low-temperature cables is based on the fact that the specific resistance of certain conductor materials is temperature-dependent (Figure 5.19).

When conventional conductors (e.g. Al, Cu) become cool, there is a notable drop in resistance that is heavily dependent on their degree of purity. At the temperature of liquid nitrogen (77 K), the specific resistance of Al and Cu with technical degrees of purity is reduced by a factor of 10 below the values at room temperature. The principle behind the so-called resistive cryocables is based on this characteristic of *normal conductors*. The combination that best lends itself as the most technically straightforward and most economical is an aluminum conductor cooled with liquid nitrogen (LN_2). For this purpose, a degree of purity of the aluminum of around 99.9 % is sufficient.

Superconductors exhibit a fundamentally different response to normal metallic conductors. Superconductors, e.g. lead or niobium, are materials whose electrical DC resistance below a *critical temperature* T_C, the so-called transition temperature, drops to immeasurably low values, i.e. practically zero. The critical temperature of most "Low Temperature SuperConductors" (LTSC) lies below 10 K; in order to achieve the superconductive state with these metals they must be cooled to around 4 K with liquid helium (LHe).

Until the mid 1980s there was no significant progress towards higher T_C values in the search for materials that become superconductive at higher temperatures. The highest

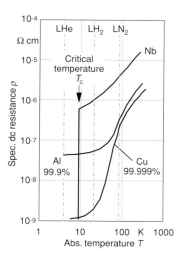

Figure 5.19
Specific resistance of normal conducting and superconducting materials

critical temperature known at present for metal superconductors is that of niobium germanium (Nb_3Ge) at around 23K/$-250\,^\circ$C (Figure 5.20). Before that time, the resistance-free and therefore loss-free transmission of current was known only in metals. The high costs involved in cooling to LHe temperatures meant that the application of metal LTSCs was limited to just a few high-tech products.

The "*High Temperature SuperConductors*" (HTSC) discovered by Bednorz and Müller in 1986 have made available materials with critical temperatures above 100 K ($-170\,^\circ$C), i.e. above the boiling temperature of liquid nitrogen. HTSCs are metal-oxide ceramics with a layered structure. In terms of system layout, the cooling of an HSTC conductor becomes considerably more reliable, more robust and more economical since nitrogen is readily available, easier and cheaper to liquefy, and is environmentally friendly.

The use of LN_2 also enables the equipment – in this case the cable – to be made far more easily and cheaply. The HTSCs therefore have far greater potential for application in the energy sector than metal superconductors. The discovery about ten years ago of high-temperature superconduction thus led the way for one of the most innovative future technologies that is currently in use in the field of cable engineering.

One of the main problems with superconductor cables, namely cooling, has come closer to solution with the HTSC. But the new oxide-ceramic materials have thrown up additional problems that must first be solved before they can be used. Table 5.9 provides an overview of the requirements for HTSC wires for industrial use in bulk-power cables [117].

Although a great deal of progress has already been made in developing technical HTSC in terms of production lengths and critical current density, there is at present still a significant gap worldwide between the requirements and the actual state of the art in engineering terms. The closing of this gap presents a great technical challenge which requires further significant R&D outlay. For equipment with relatively weak magnetic fields, e.g. bulk-power cables, the development that has already been carried out has delivered conductor properties which come so close to the specified values that the prospect of a nitrogen-cooled system looks highly promising. Based on this prospect, demonstration systems and prototypes have now been constructed using HTSC conductors.

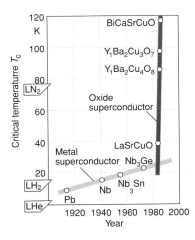

Figure 5.20
Transition temperatures of metal and ceramic superconductors plotted against time

Table 5.9
Requirements for HTSC wires for industrial use in high-capacity cables [117]

Electrical requirements	• Current density in HTSC $jc \geq 100$ kA/cm^2 (at 77 K and $B = 0.1$ T) • Currents >100 A • Low electrical losses in alternating field
Mechanical requirements	• Flexibility, manageability and sufficient mechanical strength ($\varepsilon \geq 0.2\%$) • Long wire lengths (>100 m to1 km) with homogenous properties
Costs	• Specific conductor costs ≤ 35 DM/(kA m)

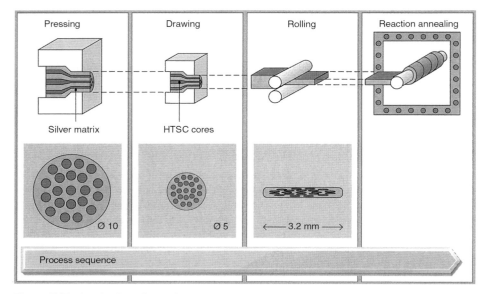

Figure 5.21 Production of multifilament silver band conductors (schematic representation)

The most promising conductor material for power cables has so far proved to be the HTSC compound of bismuth-strontium-calcium-copper-oxide (BSCCO 2223) with a critical temperature $T_C = 110$K (Figure 5.20). Because of their brittleness, these metal oxide ceramics are not suitable for forming directly into wires or strips. In comparison with the ductile metal LTSCs, the mechanical properties of these HTSCs are much less useful (*fracture elongation* $\varepsilon \approx 0.2$–0.3%).

All round the world, the most widely-used method of manufacturing wires is the *powder-in-tube method*, in which the ceramic HTSC material is poured in powder form into a silver tube and then shaped into a strip or wire through extrusion, drawing and rolling (Figure 5.21). For both mechanical and electrical reasons only multi-filament conductors are used nowadays, in which the HTSC core is divided within the silver matrix into a large number of filaments (up to 100). The current leaders in this field of wire development are Sumitomo (Japan), American Superconductors (USA) and Siemens/VAC (Germany).

The development efforts are focusing on achieving high current densities uniformly over as great a length of conductor as possible. The peak values in the SC material, depending on wire length (up to approx. 1000 m) and without any magnetic field, are currently in the region of 10 kA/cm^2 to 40 kA/cm^2 at 77K. The development aim is to increase current density for economic reasons to the range from 50 to 100 kA/cm^2. In short tests, values of 60-70 kA/cm^2 (77K) have already been achieved, and a value of 100 kA/cm^2 is generally regarded as achievable.

5.3.1 Cryocable with a Normal Conductor

Taking the force-cooled conventional cables described in Section 5.1 as the starting point, deep-cooled normal conductor cables – otherwise known as *resistive cryocables* – form the logical next step. In place of water or oil as coolants, cryogenic fluids with low boiling points would be used in cryocables. With this type of cable design, however, not only is the heat loss produced in the cable removed axially by the coolant, but at the same time the lowering of the specific DC resistance of normal metallic conductors with a high level of purity is exploited to increase the current carrying capacity and power density.

Aluminum cooled with liquid nitrogen (LN$_2$) has proved to be a good conductor material for cryocable. This has the effect of reducing the DC resistance by approximately a factor of 8 relative to normal temperature. Because of the high conductance of aluminum at the temperature of the LN$_2$, marked increases in AC resistance relative to DC resistance occur in the conductor in this type of cryocable due to current displacement (skin effect and proximity effect). The seriously reduced penetration of the alternating current relative to that at ambient temperature results in poor utilization of the conductor cross section; this factor must be addressed through appropriate conductor design (e.g. single-wire insulated segmental conductor).

As with a high-pressure oil-filled cable, the coolant – in this case LN$_2$ – can be used at the same time as an impregnating medium for electrical insulation by virtue of its excellent dielectric properties. As a solid component with the lowest possible dielectric loss factor, the use of synthetic paper is suggested for the dielectric. The tanδ of synthetic paper dielectric impregnated with liquid nitrogen is around $0.5 \cdot 10^{-3}$. Alongside lapped insulation impregnated with LN$_2$, liquid nitrogen on its own and vacuums have also been studied as alternatives.

The technical feasibility and economic viability of a cryocable are determined by the quality and dependability of its thermal insulation amongst other factors. With conventional cables the objective is to effectively dissipate to the surrounding soil any heat loss caused in the cable by achieving the lowest heat transfer resistance possible. With low-temperature cables, on the other hand, engineers are striving to achieve the exact opposite, namely the greatest possible heat transfer resistance between the soil and the cable surface. Thus steps should be taken to ensure that as little heat as possible from the soil, which is at the ambient temperature, penetrates radially inward through the cable surface to the deep-cooled conductor system (Figure 5.22). Because of the low efficiency of refrigerating machines at cryogenic temperatures, the result would otherwise be unacceptable warming of the conductor.

Various different methods of thermal insulation are employed in low temperature technology. By far the best properties are achieved with so-called *super-insulation* or *multi-*

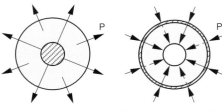

Conventional cables:	Low temperature cables:
• Radial heat transfer to the environment	• Axial heat transfer to cooling stations • Heat take-up from the environment

Figure 5.22
Heat flux in conventional underground cables and low-temperature cables

layer insulation. These terms refer to a system comprising layers of highly reflective screening material with alternate intermediate layers that are poor heat conductors under a vacuum. For example, Al foils are used with fiberglass paper as the intermediate layers. With around 20 layers of this superinsulation, between 300 K and 80 K with a good vacuum, a heat flow density of around 5 W/m^2 can be achieved. A value in this region can be regarded as sufficient for cryocable with LN$_2$ cooling [118].

Two functional areas are generally viewed separately for low-temperature cables, namely the thermal insulation and the conductor system with its electrical insulation. From the point of view of mechanical structure, three fundamentally different designs of cable have emerged:

- *Rigid cable design*
  The thermal insulation and the conductor are assembled from rigid tubes.

- *Semi-flexible cable design*
  The thermal insulation consists of rigid tubes, while the electrical conductor system is flexible (this structure corresponds to conventional pipe-type cable).

- *Flexible cable design*
  In these cables, both the conductor system and the thermal insulation (which consists of corrugated tubing) are flexible.

Figure 5.23 shows an example of a semi-flexible cryocable for three-phase alternating current. The three conductors in the cable are housed in a pressure-proof aluminum tube. In this case the coolant flows only outside the conductors ("direct surface cooling"). Versions which also have "direct conductor cooling" (using hollow conductors) have also been studied [119].

In the 1960s and on into the 1970s, these types of resistive cryocables were developed in the USA and Japan; in Germany, F&G and KFA Jülich worked together to set up a wide-ranging feasibility study [118, 119]. General Electric in the USA, for example, has developed and tested a semi-flexible cryocable with LN$_2$-impregnated synthetic paper insulation for a transmission capacity of 3000 MVA at a nominal voltage of 345 kV. Calculations based on the results suggest that cryocables should be technically capable of being used for transmission capacities up to around 3000 MVA on the 400 kV level.

However, estimates of the system costs and transmission losses have led to the conclusion that cryocables with normal conductors are inferior to conventional cables with

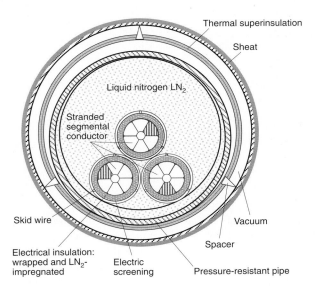

Figure 5.23
Three-phase cryocable of semi-flexible design with LN_2 cooling (schematic representation)

forced cooling for capacities up to approx. 1500 MVA, and to gas-insulated tubular conductors for higher capacities. No practical use has therefore yet been found for this type of cable construction.

5.3.2 Superconductor Cables

The use of cryogenic liquids for cooling a high voltage cable only delivers significant advantages if the conductor losses are reduced through the use of superconductors to zero or virtually zero. This characteristic of the superconductors enables the efficiency of bulk-power transmission to be increased from around 95 % to levels > 98 %.

While direct current flows in a superconductor with virtually no loss whatsoever, hysteresis losses occur under the influence of alternating fields, and as the magnetic flow density increases, so these losses increase sharply. With small flow densities (up to about 0.2 T), however, the losses are so slight that certain superconductors, e.g. niobium, can be used in three-phase AC cables. Therefore, from the outset, superconducting cables for three-phase alternating current were developed alongside superconducting DC cables. The layout design of conductors in three-phase AC superconducting cables must satisfy two criteria:

- On the surfaces of the superconductor the magnetic flow density must be kept as low as possible in order to minimize the losses.

- The electromagnetic fields must be restricted to areas between the superconductors in order to prevent eddy-current losses from occurring with the high currents in the normal-conducting components of the cable.

Table 5.10 Advantages of superconducting DC over three-phase AC high voltage cables

- Greater selection of superconducting materials for the conductor
- Simpler construction of conductor system
- No superconducting shielding, i.e. no superconducting return conductor required
- Smaller cable diameter, therefore less heat take-up from outside
- No dielectric losses
- Considerably smaller total losses (approx. 50 %), therefore less cooling performance needed
- No limit to transmission distance

The criteria for the cable can be satisfied simply by surrounding each of the conductors, which are separated by the dielectric, by another superconducting coaxial return conductor. A suitable cable layout enables a situation to be produced in which an equal current is flowing through the return conductor, but in the opposite direction from the inner conductor. This results in a virtually total field compensation. This layout does, however, have the disadvantage of its greater requirement in terms of superconducting material.

Before considering the other requirements and details of three-phase AC superconducting cables, mention should be made of the main differences between superconducting DC and three-phase AC high voltage cables.

In general terms it should be stated that superconducting DC high voltage cables offer a number of advantages over three-phase AC cables. The most important technical differences between these two constructional types are compiled in Table 5.10. This table points to the conclusion that DC superconducting cables are superior to three-phase AC cables in practically all important technical aspects. Nevertheless, as with conventional HVDC cables (Section 3.7.1), the possibilities for using superconducting DC cables for bulk-power transmission over long distances are limited because of their cost. The reason behind this lies in the converter installations that are required. It has been calculated that superconducting AC cables offer the more economical solution for transmission distances < 100 km. For this reason, development concentrated on three-phase AC cables, and therefore the following sections likewise concentrate on that technology.

Cables with metal superconductors (LTSC)

Development started in the 1960s.

The material most commonly used as the conductor for 3-phase AC cables was pure niobium, although Nb_3Sn was also studied. The advantages of niobium lie with the low AC losses, the relatively low cost and the ease with which wires and conductors can be produced. With niobium the current flows only in a thin surface layer a few mm in thickness, in other words the conductors need only a small cross section. The critical temperature T_C for niobium lies around 9 K, and can only be reached by cooling with liquid helium (LHe).

An important consideration when dimensioning the conductor is the stability in the face of transient temperature increases – e.g. as a result of short circuits – which can lead locally to a breakdown of the superconductivity. The superconductor is therefore stabilized by combining it with a normal metal offering good conductivity of electricity and heat.

The preferred stabilization materials are extremely pure copper or aluminum. In the conductors with thin niobium layers, for example, the stabilization metal serves simultaneously as a mechanical carrier and prevents the thin superconducting layer from melting – e.g. in the event of a short circuit – by removing the lost heat and creating a parallel current path. For the stabilization to function properly there must be a good metallurgical bond between the normal conductor and the superconductor.

The next of the structural components of a superconducting three-phase AC high voltage cable, the electrical insulation, needs most importantly to exhibit a low dielectric dissipation factor along with the necessary electrical stability.

In general, LTSC cables are designed in accordance with the "cold dielectric design" principle, i.e. the conductor and dielectric are maintained at a cryogenic temperature. This means that the dielectric losses of the cable from cryogenic to ambient temperature must be removed, and the cooling systems installed for that purpose are relatively inefficient.

As a rule the following dielectrics are considered for LTSC high voltage cables:

- Vacuum,

- Liquid helium (i.e. coolant same as dielectric),

- Lapped insulation impregnated with liquid helium.

Like a vacuum, LHe has practically no dielectric losses. Because of the costly support equipment that both these alternatives require, however, their use in practice is accompanied by considerable problems. For this reason, efforts were concentrated from an early stage on lapped insulation of paper or foils. Paper-like plastic foils ("synthetic paper"), impregnated with liquid helium, exhibit the best dielectric properties ($\tan\delta$ at 4 K in the range of 10^{-5}).

Along with the electrical properties of the insulation materials, their mechanical properties also have an important role. The high level of thermal shrinkage and embrittlement of plastics at low temperature pose a particular problem. Only by choosing the right material can the mechanical destruction of the lapped insulation be avoided when it is cooled down to the operating temperature of around 4 K.

The main problem with low-temperature cables – as mentioned above for cryocables – lies in the thermal insulation of the cable and the associated cooling system. In order to maintain LTSC cables at the required temperature close to absolute zero, axially flowing coolants must be used to remove the heat generated in the cable itself and the heat penetrating from outside. The hysteresis-dependent and current-dependent losses of the conductor, the dielectric losses and, under certain circumstances, eddy-current losses occurring in metal components must be taken into account as internal heat sources of the LTSC cable. However, the heat penetrating into the cable from its surroundings is of far greater significance. Along with the pressure to minimize internal loss, the requirements for thermal insulation in an LTSC cable are therefore particularly high.

The removal of heat from an LTSC cable is costly, and therefore jeopardizes the economic viability of the system, since – as mentioned above – the cooling equipment does

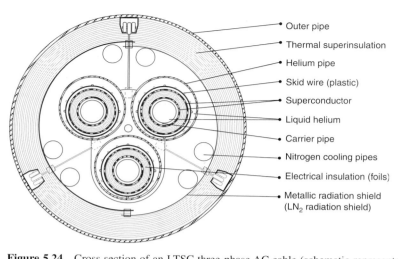

Outer pipe
Thermal superinsulation
Helium pipe
Skid wire (plastic)
Superconductor
Liquid helium
Carrier pipe
Nitrogen cooling pipes
Electrical insulation (foils)
Metallic radiation shield
(LN$_2$ radiation shield)

Figure 5.24   Cross section of an LTSC three-phase AC cable (schematic representation)

not operate efficiently at the temperatures generally encountered here. The re-cooling of 1 watt of losses using helium cooling machines with an operating temperature of 4 K requires around 350 to 500 W of electrical power. In a cryocable with nitrogen cooling machines working at 80 K, about 10 W of power are sufficient [42].

The superinsulation described under cryocables is not sufficient on its own for LTSC cables. A useful method for further reducing the heat input from radiation and conduction is the introduction of an actively cooled *intermediate shield* at a higher temperature level, from which the heat can be removed more efficiently. Most LTSC cable designs therefore provide an LN$_2$-cooled intermediate shield between the outer tube (approx. 300 K) and the helium tube (see Figure 5.24).

The thermal insulation of an LTSC cable accordingly consists of an LN$_2$-cooled radiation shield and the superinsulation in conjunction with high vacuum chambers. Only this whole package of measures can provide sufficient thermal insulation. Under realistic conditions a thermal insulation quality of < 1 W/m^2 can thus be achieved.

LTSC cable projects

Starting around 1965, great efforts began around the world to develop superconducting high voltage cables with metal superconductors. About 10 notable cable projects were initiated in Western Europe, the USA, the USSR and Japan. Along with rigid pipe configurations and fully-flexible cable types (corrugated pipe versions), semi-flexible cable types were studied as the top priority.

Of particular interest were the project run by the Brookhaven National Laboratory (BNL) in the USA , that run by Siemens in Germany, and the project run by a working partnership between Kabelmetal and the Anstalt für Tieftemperaturforschung (ATF) (Institute for Low-Temperature Research) in Austria. Table 5.11 shows certain important design and construction data from these three LTSC cable developments.

Table 5.11
Design specifications and characteristic quantities of three selected LTSC cable developments [42, 119, 120, 121]

Project run by	Siemens	ATF/Kabelmetal	BNL/USA
Voltage type	AC	AC	AC
Nominal voltage in kV	110	110	138
Capacity in MVA	2000	300	1000
Design	three-phase, semi-flexible	single-phase flexible	three-phase semi-flexible
Conductor arrangement	Coaxial	Coaxial	Coaxial
LTSC material	Nb	Nb	Nb_3Sn
Electrical insulation: Winding, impregnant	Synthetic paper with PE foil, LHe	Cellulose paper, LHe	PP foils, LHe
Thermal insulation	Superinsulation with LN_2 radiation shield	Superinsulation with LN_2 radiation shield	Superinsulation with LN_2 radiation shield
Outer pipe dia. in mm	500	220	400
Trial length in m	35	50	120
Completion of project	1978	1980	1986

The cable produced by ATF/Kabelmetal was tested in a short-term mains trial in 1979. This was the first time a superconducting cable had been tested under operating conditions in a high voltage network [42].

The superconducting 110 kV three-phase AC cable developed by Siemens [120] conforms to semi-flexible design with a thermal protection system comprising rigid tubes (see cross section, Figure 5.24). The conductor system of inner conductor and return conductor is flexible. The individual phase leads (conductors) consist of a flexible plastic carrying pipe, on which the 3-mm-diameter round niobium wires, stabilized with extremely pure Al, are wound in a spiral fashion. The thickness of the niobium layer is 50 μm, and the total niobium cross section in the conductor is around 50 mm². Over that is lapped the dielectric, composed of synthetic paper tape and polyethylene foils with a wall thickness of 14 mm. The cable is completed by the coaxially fitted return conductor. The inner conductor, return conductor and the electrical insulation impregnated with LHe are each accommodated in a helium-filled tube in accordance with the "cold dielectric design" principle.

Figures 5.25 and 5.26 show the step samples of the 110 kV phase conductor (diameter 100 mm) and of the complete LTSC three-phase AC cable of the type described here. The total diameter of this high-capacity cable amounts to only 500 mm. As in the engineering of conventional pipe-type cable, the rigid heat protection pipe system would have to be laid first, with the flexible phase conductors being drawn into the three helium tubes afterwards.

In 1977, Siemens assembled a single-phase 110 kV test system 35 m in length based on this type of cable (with terminations), and it was successfully operated in a synthetic circuit both under high voltage and with current up to 10 kA simultaneously. Using this lab-

Figure 5.25
Sample of a flexible, coaxial phase conductor (Siemens)

Figure 5.26
Model of a semi-flexible LTSC three-phase AC cable for 110 kV / 2000 MVA (Siemens)

oratory test run in Erlangen it was finally possible to demonstrate the technical feasibility and the capacity of LTSC cables with helium cooling. However, economic considerations and system management aspects have not led to a situation in which this type of cable has been installed for practical applications. The project was abandoned in 1978.

The results obtained in the other LTSC cable projects also confirmed that the underground transmission of power using metal superconductor cables is fundamentally possible from a technical point of view. Without doubt the technical high point was achieved in the Brookhaven National Laboratory (BNL)/USA project, which continued right through to 1986, i.e. shortly before the discovery of HTSC [121].

In all the application-oriented test objects the main disadvantage emerged as the extremely high cost of conductor cooling and the thermal insulation, i.e. the thermal protection system. As a result of these factors, the specific investment and transmission costs only fall below those for conventional force-cooled cables with extremely high transmission capacities. Studies and estimates of the "break even point" for cost-efficiency carried out in the mid 1970s led to the conclusion that the transmission capacity of LTSC three-phase AC cables would need to be 4000 to 5000 MVA per system before that point would be reached [122]. Such high-performing systems were then – and still are today – way in advance of reality. The development of superconducting cables with metal superconductors (LTSC) was therefore shelved worldwide when the three-phase AC cable project at US BNL finished in 1986 [121].

So here was another type of low-temperature cable that had failed to emerge as a realistic alternative for bulk-power underground transmission.

Cables with oxide-ceramic superconductors (HTSC)

With the discovery of the high-temperature superconductors (HTSC) with critical temperatures >90 K, the application of superconduction in power cable engineering returned as the central international focus of scientific and technical interest. The appeal of the new oxide-ceramic-based superconductors lies especially in their high critical temperatures, which allow operation at 77 K, the boiling point of liquid nitrogen. These developments have led to simpler, cheaper cooling systems (approx. factor 15:1) operating far more efficiently (approx. factor 40:1) and simpler, cheaper thermal insulation systems [124]. Furthermore, LN_2 is much cheaper than LHe, and is available in limitless quantities.

As can be seen from Figure 5.27, switching from the proven LTSC cable ("*4-K cable*") to an HTSC cable ("*77-K cable*") results in a considerably simpler configuration and a notably smaller cryosheath, and therefore in a reduction of external diameter from 500 to approx. 350 mm. In its basic construction the HTSC cable is similar to the normal-conducting cryocable. The thermal losses from the HTSC cable are roughly three times lower than from the "4-K cable". If the lower cost of the cooling stations and coolant are also taken into account, then assuming that the phase conductors cost the same as in LTSC cables, and assuming the same transmission capacity, this results in a reduction in investment cost of around 25%.

These advantages of the HTSC cable ultimately lead to lower system and operating costs relative to LTSC cables. The above assertions will only hold true, however, if the technical capacity is comparable and if the cost of the HT superconductors comes down to the same as or less than the cost of LT superconductors. Thus, using HTSC cables, economically viable power transmission could be achieved even at capacities ≤1000 MVA. The criteria that must be satisfied are shown in Table 5.9 (see above).

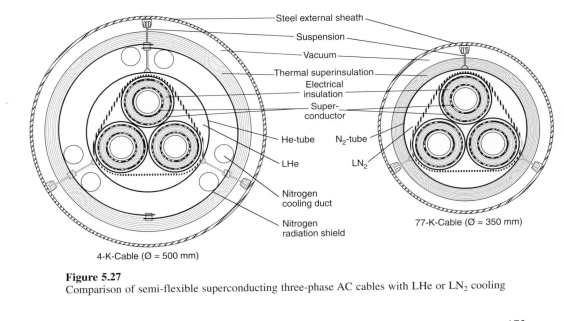

Figure 5.27
Comparison of semi-flexible superconducting three-phase AC cables with LHe or LN_2 cooling

Because of the fact that, as things stand today, the conductor costs amount to a very high proportion of the overall costs of the HTSC cable, the forecasting and implementation of future conductor prices is of great importance for the economic competitiveness of this cable. The production costs of the BSCCO strip conductor in the so-called "*oxide powder in tube*" (*OPIT*) process (Figure 5.21) are currently influenced predominantly by the forming of the tubular conductor and the thermo-mechanical operations, i.e. the number of steps involved. Other important factors for reducing the specific costs are increasing the current density and the fill factor. The cost of the SC powder is currently fairly insignificant in this respect. Only through considerable simplification and rationalization in the manufacture process of the wires or strips will it be possible to achieve the development target for conductor prices.

A basic requirement for using HTSC in the field of power cables, however, is the achievement of a *critical current density* j_c in the superconductor sheath of approx. 100 kA/cm^2. With an HTSC content of 25 to 30% by volume, this results in technically relevant current densities j_c over the entire cross section of the conductor ("engineering current density") of around 20 to 30 kA/cm^2 [117]. As explained at the outset in Section 5.3, that situation is still some way in the future. Increasing the critical current density thus represents one of the main tasks for development projects today and in the future.

Amongst other things, these objectives led six leading European cable manufacturers (including BICC, Pirelli and Siemens) as early as 1990 to collaborate in starting and progressing a pre-competitive research program in the HTSC field. This first joint industrial project in this field in Europe was supported by the European Community in the framework of the BRITE-EURAM I, II and JOULE programs. This work was successfully concluded in 1995. The results achieved lent encouragement to specific single-company R&D projects in the HTSC cable field and led Siemens and Pirelli to invest further effort in this new technology for bulk-power underground transmission.

A technical/economic study has been carried out based on the preliminary work, and this has shown that HTSC cables should be capable of bulk-power transmission (up to about 1000 MVA) on the 110 kV level in a way that is more advantageous than using force-cooled 400 kV oil-filled cables. In certain cases, bulk-power transmission using these 110 kV HTSC cables may even mean that it is possible to operate without the 400 kV level, i.e. the number of voltage levels and therefore the number of power transformers could be reduced. On condition that the development targets and requirements for the HTSC cable conductor in terms of current density and specific conductor costs (Table 5.9) are met, then the investment costs relative to conventional cables of equivalent capacity would fall by around 30%, and the transmission losses by around 80%.

Figure 5.28 shows a water-cooled 400 kV oil-filled cable system and the design of the 110 kV HTSC cable system for comparison. The advantages of the 110 kV HTSC three-phase AC cable in terms of its space requirement, weight and losses are considerable. Table 5.12 shows examples of possible applications for HTSC cables in existing three-phase AC networks. Based on the economic aspects, it will only be worth using these cables in situations where the saving of losses is regarded as very important [125]. From where we stand today, the applications are initially limited to fields with a high power density for each route, where a second cable system needs to be provided in order to safeguard transmission.

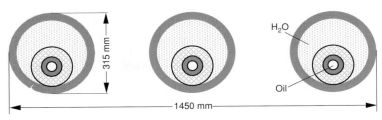

a) 380-kV-LPOF cable, directly surface-cooled by water
    (Bewag I und II, Berlin)

    Weight of cable core   28 kg/m
    Total losses   190 kW/km

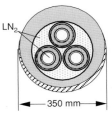

b) 110-kV-HTSC cable, nitrogen-cooled

    Weight of core 7 kg/m
    Losses   37 kW/km

Figure 5.28   Comparison of force-cooled bulk-power cables for 1000 MVA transmission capacity

Table 5.12   Application possibilities for HTSC cables in three-phase AC networks

Field of app'n	Objective	Voltage in kV	Line length in km	Power range in MVA
Supply in conurbations	Replacement of 400 kV-cables and removal of need for transformers in cities	110	≤ 20	1000–1500
Distribution in conurbations	Replacement of conventional cables in in the event of routing problems	60–110	≤ 10	300–500
Outgoing lines from power stations	Outgoing power line at generator voltage and removal of need for transformers in underground power stations	20–30	≤ 1	300–700
Bulk-power transmission	Alternatives to overhead lines on partial sections (e.g. for environmental reasons)	110 (400)	≤ 20	≥ 1000

The installation of a power supply system with significant area coverage based purely on HTSC equipment appears unrealistic in the long term according to the stage of development at present.

HTSC cable projects

Development of HTSC cables is being carried out primarily in Japan (Furukawa, Sumitomo), in the USA (Pirelli Cable Corporation, Southwire) and in Europe (Pirelli, NKT, BICC).

In all of these cases HTSC wires or strips made from BSCCO 2223 with a silver matrix are used, and liquid nitrogen is employed as a coolant. The semi-flexible construction is used in preference, and both the cable that is only internally cooled with a "warm" dielectric at ambient temperature and the coaxial type of cable with "cold" dielectric at LN$_2$ temperature are being studied. The first type is specifically designed for the "retrofitting" of high-pressure oil-filled cables in a steel pipe (pipe diameter approx. 200 mm). The advantage of this design lies in the lower costs (only half as much HTSC material is required), while its higher losses represent a disadvantage. The capacity at 115 kV is therefore limited to 400 MVA. The Pirelli Cable Corporation, ASC and the Electric Power Research Institute (EPRI) are all involved in this project, which is aimed specifically at the US market.

The most firmly established projects are being conducted under the leadership of the Tokyo Electric Power Company in Japan, in collaboration with Furukawa and Sumitomo. The goal of their development is to assemble and successfully trial a 66 kV three-phase AC cable for 1000 MVA, with the intention of using it in place of conventional cable. Prototype cables several meters in length, including the necessary sealing ends, have already been produced.

Pirelli is developing a 110 kV HTSC three-phase AC cable for 400 to 1000 MVA. Figure 5.29 shows a draft design for this semi-flexible three-phase cable. Current-carrying capacities of 2000 A were reliably recorded in the first functional models with flexible conductors. The coaxially constructed phase conductors with a diameter of 125 mm have

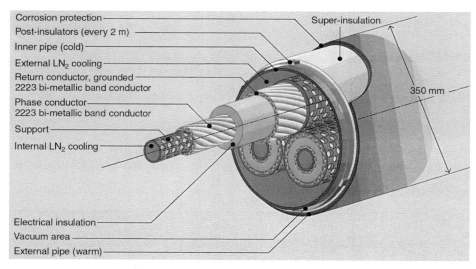

Figure 5.29
Design of a 1000 MVA/110 kV three-phase AC cable with high-temperature superconductors

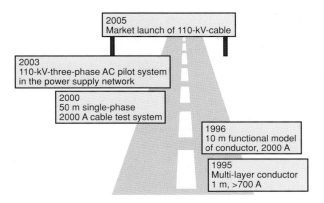

Figure 5.30 Road map for the development and testing of a 110 kV HTSC cable

Table 5.13 Problems to be solved before HTSC cables are actually installed

HTSC wires	• Increasing critical current density • Improving mechanical strength • Manufacturing wire reproducibly for long lengths • Reducing the costs of the wire
HTSC conductors	• Low electrical losses with alternating current • Behavior in the event of a short circuit
Dielectric	• Electrical strength of LN_2-impregnated insulation • Reducing the dielectric losses • The effect of bubbles in the LN_2
Thermal insulation (cable cryostat)	• Reducing the thermal losses • Mechanical behavior when pulling in the phase conductor • Continuous production procedures • Contraction when cooling down • Reducing the costs
Cooling system	• High-performance cryo-coolers • Improving the efficiency • Freedom from most maintenance tasks
Accessories	• Dielectric and thermal dimensioning • Wire connection techniques • Contraction during cooling
General	• Cable laying • Repair techniques • Redundancy criteria

an LN_2-impregnated insulation layer comprising PE foils and synthetic paper 12 mm in thickness.

Preparations are being made to produce a single-phase prototype cable 50 m in length as the next step. Most cable projects plan to conclude their development phase with a laboratory trial of single- or three-phase prototype cables before the end of the century. The network trial of pilot systems is planned for the beginning of the next century. It is not anticipated that the HTSC cables will be ready for commercial applications before 2006. Using the above mentioned 110 kV HTSC cable project as an example, Figure 5.30 shows in the form of a "road map" the main development milestones.

At the IEA workshop on the subject of "HTSC power transmission cables" held in Milan in April 1997 the questions yet to be solved and the existing problems were discussed by participants from all the project groups. Table 5.13 lists the main points covered by that discussion. It is clear from this that considerable development effort is still needed before the HTSC cables will be ready for practical applications.

5.4 Technical, Economic and Ecological comparisons Between the Various Systems

In the field of bulk-power transmission, overhead lines take precedence over underground cables because of the significantly lower costs as well as for technical reasons. In terms of technical developments as they stand today, the overhead line is a powerful, operationally reliable and economical transmission system. All the technical problems associated with the equipment have largely been solved. Extra high voltage overhead lines are used predominantly as overland connections; in conurbations it is difficult to find suitable routes for them. In the future, therefore, it will not generally be possible to advance into densely-populated areas with overhead lines [126]. Environmental compatibility factors (e.g. "visible" equipment or Electromagnetic Compatibility EMC) are also making it increasingly difficult to install overhead lines. Suitable types of cable will be in a position to exploit this situation. The greatest challenge and opportunity for underground transmission systems is the job of supplying large cities with a high load density.

Tried-and-tested methods are available for carrying out these tasks with low-pressure oil-filled cables and XLPE cables. The range of transmission capacities in naturally cooled cables at 110 kV can reach around 200 MVA, and at 400 kV it can reach around 800 MVA per system (Figure 3.37). For higher capacities, indirect and direct cooling of the cable surface with water have become the established methods, and they have proved their worth in operation over a period of decades. Transmission capacities can thus be increased to levels up to 500 MVA (110 kV) or 1500 MVA (400 kV), which is sufficient for most of the tasks and projects that are under discussion.

The direct cooling of the cable conductor (with water or oil) enables a further increase in capacity. Nevertheless, development of this transmission system was shelved because the technical cost was too great and because it was not economically viable enough.

For transmission currents above 2 kA, i.e. around 400 MVA at 110 kV or around 1500 MVA at 400 kV, gas-insulated lines (GIL) and low-temperature cables lend themselves for consideration as an alternative.

As regards GIL, so far only the rigid version has become established in lengths < 1000 m and laid in air (e.g. in tunnels or open trenches). Transmission capacities of 3000 MVA are possible and have already been achieved. In the capacity range < 1500 MVA and at voltages < 275 kV the economic viability of GILs is questionable in comparison with force-cooled cables. Further developments towards underground installation or installation in tunnels and a reduction in cost could mean that this technology might qualify as an underground alternative to overhead lines in the voltage range above 275 kV. Flexible SF_6-insulated cables, on the other hand, were unable to gain acceptance for technical and production-related reasons. Those developments have been abandoned.

As regards low-temperature cables, the development of cryocables and superconductor cables with metal superconductors was eventually shelved due to poor economic viability.

The high-temperature superconductor cable is regarded as a possible alternative to naturally-cooled and force-cooled cables in the capacity range from about 400 to 1000 MVA. But the technical feasibility and – most importantly – the economic viability of this new technology must first be demonstrated. It is debatable whether HTSC cables could one day also be used in place of overhead lines for higher capacities. Table 5.14 provides a

Table 5.14   Comparison between various three-phase AC bulk-power transmission systems

System	Development status[1]				1997 status	Problems	Environmental aspects
	1	2	3	4			
Naturally cooled cable up to 500 kV (oil-filled and XLPE)	✓	✓	✓	✓	Proven technology	Limited transmission capacity	"No oil" problem
Cable with indirect and direct surface cooling using water	✓	✓	✓	✓	Proven technology	Techn. resource demand, econ. viability	"No oil" problem, High losses
Cable with direct conductor cooling using water	✓	✓			Development discontinued	Techn. resource demand, econ. viability	problem,
Rigid GIL in air, up to 500 kV (<1000 m)	✓	✓	✓	✓	Proven technology	Underground installation, econ. viability	SF_6 insulating gas
Flexible SF_6 insulated cable up to 345 kV	✓	✓			Development discontinued	Tech. resource demand, manufacturing	
Rigid GIL underground or in air up to 500 kV (>1000 m)	✓	✓	⇨		Potential exists	Insulating gas, economic-viability	SF_6 insulating gas, space requirement
Cryocable with normal conductors	✓	✓			Development discontinued	Economic viability	
Superconductor cable with metallic super-conductors (LTSC)	✓	✓			Development discontinued	Techn. resource demand, econ. viability	
Superconductor cable with ceramic super-conductors (HTSC)	✓	✓	⇨		Potential exists	HTSC wires, Economic viability	
Overhead lines up to 500 kV	✓	✓	✓	✓	Proven technology	Largely solved	"Visible", Space required EMC

[1] 1: Feasibility studies   2: R&D prototype systems   3: Pilot systems in supply netwok   4: Operational use
✓ finished
⇨ in progress

comparison between the various aspects of the different three-phase AC bulk-power transmission systems.

It should be noted here that the basic requirements for underground power transmission have also changed significantly over the last decade. While, in the 1970s and 1980s, interest focused primarily on increasing transmission capacity, in the last few years the cost and environmental aspects have taken center stage. So the most pressing problems for underground bulk-power transmission today – with transmission capacities often lower than before – are the reduction of investment and operating costs, the reliability, and thus the guarantee of successful transmission.

From the ecological point of view, attention is focused mainly on the subjects of landscape protection ("invisible" systems by laying them underground), the replacement of low-pressure oil-filled cables with extruded cables (general principle of "no oil" in the cable), reducing losses and space requirement (e.g. narrow routes) and finally EMC. Attention should also be paid to environmental factors when selecting materials (no oil to pose a threat to water supplies, no PVC, no large amounts of SF_6), the service life and life cycle costs of the cables, and their recyclability.

There is now a solution for underground bulk-power systems up to 500 kV in the form of XLPE-insulated cables, and this solution is also quite satisfactory in terms of environmental compatibility. This solution satisfies the criteria of "no oil" and reduced losses (through smaller $\tan\delta$). Forced cooling is possible with this single-conductor cable type, as with the low-pressure oil-filled cable, and this in turn leads to increased capacity. In comparison with overhead lines, the EMC problems with underground cables are smaller from the outset, and can be further reduced by the use of metal sheaths, optimized layout and, if appropriate, the introduction of compensating conductors (Section 7.3).

In the case of gas-insulated lines (GIL), the main issues to be addressed from the point of environmental friendliness are the wholesale replacement of SF_6 by a different insulating gas (e.g. nitrogen) and the possibility of underground installation.

With the latest technology – HTSC cables – many ecological aspects were taken into account from the very start at the design stage. Most designs provide for three-phase arrangements in a common metal pipe, with liquid nitrogen being used as a coolant and impregnating medium. The losses from the cable are slight and are not dissipated into the surrounding soil, and the dimensions are comparatively small.

During the planning and design stages of underground bulk-power systems, issues of environmental compatibility generally need to be addressed nowadays along with the technical and economic comparisons. However, the so-called environmental compatibility test adds a further level of complexity to the process of decision making. The resulting restrictions and conditions for the construction and operation of bulk-power systems can and will – as demonstrated in the past – lead to innovations in the field of underground transmission technology in the future. An example of this is the 400 kV diagonal connection installed by Bewag in Berlin [127].

Only with the boundary conditions that occur in individual cases can a comparison and evaluation ultimately be carried and the ideal solution chosen and implemented.

6 Cable Accessories

High voltage and extra high voltage cables can be produced in relatively long continuous lengths of several km, and therefore in theory can be adapted for extended routes (see Sections 4.1.4 and 4.2.3). Laying such cable "in sections", however, is generally not possible due to the problem of transportation. Drums must not exceed specified maximum dimensions if they are to be transported by the manufacturer to the site by rail and road. Thus, the maximum size of cable drums – and by implication the greatest lengths that can be transported overland – are determined most significantly by the rail and road profile. Submarine cables form an exception; they are generally loaded onto ships by the manufacturer and dispatched in one-piece lengths ready to lay.

Depending on the conductor cross section, the rated voltage and the cable construction, delivered lengths of between several hundred meters and approx. 2 km can be achieved in the high voltage and extra high voltage field. Consequently, in order to put together longer sections of cable from limited individual lengths, specialized connection components are required: *joints*. Where cables of the same type are to be joined together, *normal straight joints* are sufficient in contrast to so-called *transition joints*, which serve to connect cables with an extruded dielectric to those with paper insulation. Furthermore, it is always necessary to electrically "terminate" the cable end with suitable components, and this function is performed by the *sealing ends*. The general term *cable accessories* is used to describe these component parts of a cable system.

Quite apart from their purely technical functions, cable accessories are important by virtue of the fact that the overwhelming majority of all faults that occur in a cable system both under operating conditions and in tests occur locally in sealing ends and joints. A major study of cable users and manufactures in Japan into the frequency with which faults occur in that country's 66 kV and 77 kV XLPE cable networks produced the result that only 11 % of the faults were attributable to cable breakdown, while 89 % were due to faults with cable accessories [128] (see Section 8.3.3., Figure 8.21). An even clearer result emerged in the course of a year-long pre-qualification test for 400 kV XLPE cable systems led by a German electricity supply company between 1994 and 1995, and with six of the leading European cable manufacturers participating [129]: All nine of the faults that arose in the course of testing were attributed to the accessories, of which six were attributed to joints and three to sealing ends. It follows that in the realm of accessories, and particularly in XLPE cable systems, there is still room for improvement.

Not least as a result of these considerations, the Japanese have been making efforts over the last few years to implement so-called *super long-length cables* in order to minimize the number of straight joints and thus to increase the reliability of the cable systems. For this purpose, drums 11.5 m wide (with a flange diameter of just under 4 m and weighing up to 100 t) are provided for overland transportation, and for river and canal transportation cable containers mounted on turntables with a 7 m diameter. In this way, 275 kV XLPE cable with 2500 mm^2 Cu conductor can be delivered in lengths up to 1800 m for laying, while the 500 kV versions can be delivered in lengths up to 1600 m.

6.1 Necessity for and Tasks of Accessories

Irrespective of the type of cable and accessories, their main tasks – as with the cable – lie in transmitting the conductor current and in providing insulation between the conductor and the ground potential under all operating conditions that may arise in the network – including test conditions and short-circuit conditions. They must perform these tasks without allowing any of the system components to sustain any lasting damage, e.g. through local overheating. Furthermore, under all types of operating conditions – including situations where faults occur – protection against electric shock must be guaranteed in accordance with DIN VDE 0105.

The sealing ends are also responsible for functioning as an *interface* between the cable system and other electrical supply systems, e.g. switchgear and transformers. This interface must ensure a clearly-defined division between the two systems, which amongst other things clearly delineates the responsibilities of the respective manufacturers.

In gas pressure cable systems with nitrogen at a pressure of approx. 15 bar (which is necessary to safeguard their characteristics in the long term), this criterion is self-evident, since neither switchgear nor transformers are designed mechanically or electrically for such conditions. Moreover, avoiding gas losses is in the interest of the operational safety of the cable system, and so its components are always designed for optimum pressure maintenance.

With XLPE cables, on the other hand, it is quite feasible from a technical point of view to integrate the components used as field control components in the vicinity of the sealing end, depending on their design and material, into the neighboring switchgear or transformer systems. For example, where a cable enters SF_6-insulated switchgear, its insulating gas could be used to embed the field control elements without involving any additional supply or monitoring systems. But in practice this technique has only been accepted to a limited extent, because combining functions in this manner makes it impossible to unambiguously define the interface between the various different items of apparatus. The possibility of gas losses, an important criterion for the quality of SF_6 installations, would no longer be restricted to the components that could be calculated by the producer of the system.

As a result, at least in the 110 kV cable systems installed in the Federal Republic of Germany, the sealing ends in each case provide a definite interface to neighboring systems. This involves gas-tight and pressure-resistant cast-resin insulators with overall dimensions and connection dimensions conforming to the international standards for switchgear.

One particularly sophisticated function of the accessories results from the fact that the cylindrically symmetrical field distribution found within the radial field cable at the connection points and ends is severely disrupted, associated with considerable *field enhancements*. The accessories are therefore assigned the task of restoring a calculable potential distribution through appropriate *field control*, the electrical loading not being permitted to exceed the long-term permitted limits for the dielectrics concerned at any point.

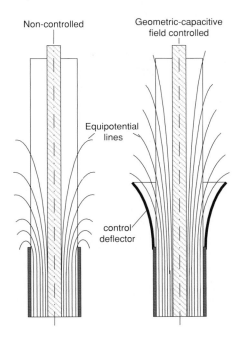

Non-controlled

Geometric-capacitive
field controlled

Equipotential
lines

control
deflector

Figure 6.7
Principle of geometric-capacitive field control

- *Geometric-capacitive field control*

  In geometric-capacitive field control (Figure 6.7) the volume capacitance C_V at the end of the conductive layer is greatly reduced by emphatically lengthening and widening the outer conductive layer in the shape of a so-called *control funnel* or *control deflector*. As defined in Formula 6.1, this results in an increase in the creepage inception voltage U_a. The deflector, made from an elastic conductive compound based on silicone rubber or EPDM, is generally integrated into a similarly elastic insulator, and the two components together form a prefabricated *stress cone*. This results in the field being moved simultaneously into the solid insulation material; the particularly critical sliding arrangement cable insulation/air no longer plays a part.

  Prefabricated stress cones with geometric-capacitive stress control were initially developed for medium voltage cables; now they represent the standard solution for sealing ends on all types of polymer-insulated cables and for all voltage levels up to 500 kV. Unlike capacitive control with concentric conductive layers, the effect that the deflectors have on the field distribution at the cable end depends to a certain extent on environmental factors (partially "open" field pattern). Therefore the optimum mounting position for the control element must be determined and, if necessary, checked for each type of sealing end.

The improvement of geometric-capacitive field control for extra high-voltage applications was only made possible by the logical use of mathematical field calculation methods. With the help of such methods it was possible to limit the overall dimensions of the stress cones for sealing ends and joints (Section 6.2.3) whilst maintaining maximum field strengths that the system can demonstrably support long-term at levels that can be controlled during manufacture and installation.

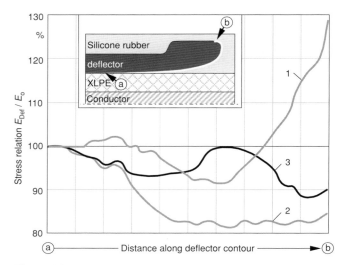

Figure 6.8
Field strength pattern along the development of different stress cone contours, as the result of numerical field calculations [131]

A material is being used to the maximum advantage in electrical terms when the field strength is consistent right across the surface of the deflector and corresponds to the value of the undisturbed cylindrical field on the outer conductive layer of the cable. Exceeding the value of this limit locally conceals the danger of electrical ageing, while falling below the optimum value has the effect of increasing the dimensions of the stress cone.

As an example, Figure 6.8 shows three characteristic E_{def}/E_O loading operations along the deflector contour of a sealing-end stress cone for 400 kV, as they were determined in the context of a large number of calculations using the simulated charging method [131] (E_{def} = field strength at deflector; E_O = value of the cylinder field at the outer conductive layer of the cable):

- Curve 1 exceeds the specified maximum field strength E_O on the front side of the deflector by more than 20%, combined with the danger of accelerated aging.

- Curve 2 fails to take sufficient advantage of the insulation, remaining virtually 20% below the permissible field strength over long sections of the deflector contour; the associated stress cone is unnecessarily long

- Curve 3 comes closest to the optimum solution; at no point is the maximum field strength exceeded, while on the other hand well over 90% is utilized over virtually the whole contour.

6.2 Designs of Sealing Ends and Straight Joints

This subsection describes accessories that perform the standard tasks of "connecting conductors", "insulating" and "field control" as described in Section 6.1. Joints with special and supplementary functions such as separating the screen (*separating joints*), connecting cables with different dielectrics (*transition joints*) and dividing the supply of impregnating medium to low-pressure oil-filled cables into several sections separated hydraulically from one another *(stop joints)* are covered separately in Section 6.3.

6.2.1 Techniques for Joining Conductors

The primary task of straight joints is to create a long-term connection between two cables of the same type; the connection must be capable of carrying current under a specified voltage between conductor and screen or metal sheath. A number of different techniques is available for this purpose and their suitability for a specific case depends, amongst other factors, on the type, material and cross section of the conductor, on the cable type, the dielectric and the nominal voltage.

Requirements for conductor connections

Irrespective of how they are implemented, connections between conductors must satisfy certain minimum requirements, particularly in respect of current carrying capacity and mechanical strength; these requirements are specified in the relevant standards (e.g. VDE 0220):

- Load-carrying capacity for the rated and maximum permissible short-circuit current, where the joint must not heat up to a greater extent than the conductor in the non-disturbed area of the cable, i.e. with the smallest transfer resistance possible. Efforts should even be made to achieve a lower level of heating through ohmic losses than that occurring elsewhere in the conductor because, otherwise, unacceptably high temperatures may develop locally due to the fact that, in the area of the accessory, heat dissipation is normally more difficult (due to thicker insulation). See Sections 3.5.1 and 5.1.2, Figure 5.5)

- Sufficient mechanical strength to withstand stresses during assembly, the stresses of temperature fluctuations and short circuits that may occur during operation.

- There are further requirements that relate mainly to the cost aspect, such as simple, quick and reliable installation under the more difficult conditions on construction sites.

Types of conductor connection

A fundamental distinction needs to be made between conductor connections that are formed thermally (by soldering and welding) and mechanically (by pressing). Both these types of connection can be used for sealing ends and joints, for plastic and paper-insulated cables with or without hollow conductors. Depending on the technique used, either *equal-diameter* or *larger-diameter connections* are produced, a criterion that affects the field distribution in joints (as shown in Figure 6.9) and is therefore of particular significance for extra high-voltage cables.

Table 6.1, in conjunction with the explanations that follow, provides an overview of the various conductor connection techniques for high-voltage and extra high-voltage cables,

189

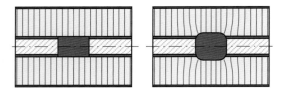

Figure 6.9
Field distribution with equal-diameter and larger-diameter conductor connections in joints (schematic representation)

Table 6.1 Conductor connection techniques for high voltage and extra high voltage cables

Category	Process	Application	Characteristics
Produced thermally	Soldering with sleeves	Cu conductors for paper-insulated cables	larger or equal diameter
	Welding	Al cond. of paper cables w/o hollow chan.	always equal diameter; too much heat for XLPE; not used in Germany for HV
	Cadweld	all conductors w/o hollow chan.	always larger diam. connection; suitable for XLPE
Produced mechanically	Hexagon-crimp	all conductors	protective tube required to protect six-sided crimp, therefore always larger diam.
	Round crimp	all conductors	always equal diameter

their advantages and disadvantages and the fields of application for which they are appropriate:

- *Soldered connections* using soldering sleeves are used mainly for Cu conductors. The sleeve can be either solid or hollow (for oil-filled cables), and may be equal in diameter or wider (Figure 6.10). With the equal-diameter version, the outer strands of the multi-wire conductor should be cut back in several steps. This is the preferred technique, especially with extra high-voltage cables, since the field disturbance for the joint insulation is less pronounced than with larger diameter sleeves. The connection is normally soft-soldered at $\vartheta < 450\,°C$ in order to limit the thermal stress on the adjoining insulation. However, if there is a possibility that short-circuit temperatures of above $160\,°C$ may develop, then the connection must be hard-soldered.

- *Welding* is preferable for Al conductors in paper-insulated cables, although these are very rarely used in the high-voltage field. Since sleeves are not required for welding, this technique always produces conductor joints with an even diameter. Conventional welded connections are not appropriate for PE or XLPE, however, because of the high temperatures that are applied for a long period.

- *Cadweld* using prefabricated shapes and powder metal, on the other hand, is also suitable for synthetic insulation as the operation lasts only a few seconds and therefore a relatively small amount of heat is developed. This technique can be used with both Al and Cu conductors, and proves advantageous particularly with large cross sections because of its excellent thermal and mechanical capability. On the other hand, it does have a drawback when used for extra high-voltage cables in that equal-diameter connections cannot be produced.

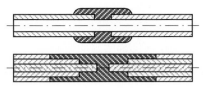

Figure 6.10
Soldered connections using a sleeve, larger-diameter and equal-diameter versions

Wherever possible nowadays (and especially at the "lower" end of the high-voltage field and where the conductors used are not too large), conductors are connected by mechanical compression, as this technique involves no thermal stress for the insulating material. With this method, hydraulic tools are used to squeeze suitable *crimp sleeves* together with the opposing conductor ends. The sleeve material matches the conductor concerned, thus Cu sleeves are used with copper conductors, Al sleeves with aluminum conductors etc. A distinction is made between two compression techniques:

- *Hexagon crimping*

  The prefabricated, originally round crimp sleeve, which is matched to the diameter of the conductor, is here shaped by the hydraulic tool into a hexagon; the intensive reshaping of the material results in very solid and conductive, yet essentially larger-diameter connections. In order to avoid dangerously enhanced fields at the pressed edges, the sleeves are covered with an additional smooth protective tube that is pushed onto the conductor in advance (Figure 6.11); in the case of oil-filled cables a steel tube with an axial drilling is provided.

- *Round crimping*

  The drawbacks with hexagon crimping arising from sharp-edged reshaping of the sleeves, and the resulting requirement for a larger-diameter protective cover are avoided by the use of special tools for round crimping. The connections are generally arranged in such a way as to have the same diameter as the conductor, and are thus advantageous for the field distribution in the area around the joint. As with equal-diameter soldering sleeves (Figure 6.10), the outer conductor wires must be cut back in steps for this purpose. Sleeves for oil-filled cables have an axial drill hole.

The so-called *plug-in systems* occupy a special place amongst conductor connection techniques; these systems were developed originally for medium voltage applications, but meanwhile have also been improved for entry sealing ends in switchgear and transformers for polymer-insulated high-voltage cables up to 150 kV. First, a sealing end comprising prefabricated components (see Section 6.2.3) is fitted, and then a laminated contact capa-

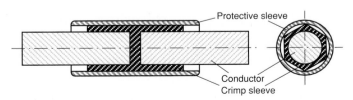

Protective sleeve

Conductor
Crimp sleeve

Figure 6.11   Hexagon crimp with a field-smoothing protective sleeve over the crimped connector

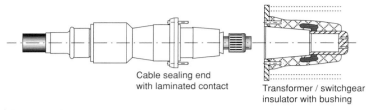

Cable sealing end
with laminated contact

Transformer / switchgear
insulator with bushing

Figure 6.12
Plug-in system for polymer-insulated high voltage cable entry sealing ends for switchgear and
transformers (schematic representation, in accordance with ABB, Pfisterer)

ble of high current is welded to the conductor. The dimensions of this contact are matched
to an integral coupling socket in the entry insulator of the switchgear or transformer (Figure 6.12). The actual conductor connection is made by plugging the sealing end into the
socket and securing the contact (for further details see Section 6.2.3).

6.2.2 Accessories for Paper-Insulated Cables

The insulation of the associated accessories matches the composition of the cable insulation, and likewise consists essentially of impregnated paper, i.e. *condenser "cones"* as
shown in Figure 6.5 for field control in the area around the sealing end and wound strips
of paper (mainly in the form of so-called *wide-web papers*) as joint insulation.

Sealing ends

Capacitive field control within the cable sealing ends is achieved by condensor "cones"
with conductive aluminum foil control layers matched in number and size to the voltage
level concerned, e.g. nine inserts for 110 kV and 25 inserts for 400 kV. However, the
precise number varies from one manufacturer to the next.

Condenser cones are produced manually from pre-impregnated paper strips up to 1.5 m
in width, either in the manufacturer's factory or on-site. On site the paper strips and conductive control layers are wound directly onto the cable insulations exposed by the outer,
field-smoothing layer, while impregnated hard paper tubes are used as mounts for the
factory-produced condensor cones. The internal diameter of the tubes is matched to the
corresponding cable conductors, and therefore depends on the conductor cross section as
well as the insulation thickness (voltage rating). The tube diameters exceed the insulation
diameter by approx. 2 mm, resulting in a gap of approx. 1 mm. This must first be
matched on site to the actual conductor dimensions by means of so-called *compensation
winding*, before the condensor cone is pressed onto the cable end and positioned precisely.

The advantage of prefabrication lies in the fact that, after production, the sealing end
insulation can be vacuum-treated effectively, and then later only comes into contact with
air once more briefly during the pushing-on process and subsequent fitting of the insulation. The prefabricated cones are stored and also dispatched to the site in sealed, liquid-
filled containers.

Winding the sealing end on site does necessitate a more thorough final vacuum treatment
because of the longer contact of the insulation with the atmosphere; more useful, on the

other hand, is the fact that the condensor cone can be matched perfectly to the actual contour of the core. This advantage comes into its own particularly with external gas pressure cables, whose slightly oval conductor shape (Figure 4.20) renders it necessary in any case for the sealing end winding to be carried out on site.

To protect the capacitor cone against moisture and mechanical damage, and more importantly to seal off the various media surrounding it (depending on where the sealing end is installed)

- Transformer oil in the case of transformer entries,
- SF_6 in the case of metal-enclosed switchgear and
- Air in the case of outdoor sealing ends,

the control cone is generally fitted in a stable insulator. For outdoor use it is made from glazed porcelain or fiber-reinforced cast resin with hydrophobic, UV-light-resistant elastomer shields (so-called *composite insulators* as shown in Figure 6.13), and when used as a transformer entry or switchgear entry it is made from cast resin. In order to make transformers and switchgear more compatible, international standards have been produced for the main dimensions of these cast-resin insulators for the various voltage ratings (IEC 859).

While the solid component of the sealing ends for paper-insulated cables is generally cellulose, the type of impregnating medium depends on the type of cable that the accessory is to be attached to:

- Low-viscosity insulating fluid for low-pressure oil-filled cables and
- High-viscosity masses (or compounds) for gas pressure cables.

Once the installation operation is complete, the sealing ends are vacuum treated again in order to remove any air and moisture that may have penetrated the insulation. They are then filled with the appropriate impregnating medium and, if necessary, positive pressure is applied. Figure 6.14 shows the cross sections of an outdoor sealing end and a sealing

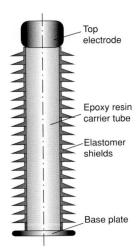

Top electrode

Epoxy resin carrier tube

Elastomer shields

Base plate

Figure 6.13
Composite insulator for outdoor installation as an alternative to porcelain

193

end for a switchgear entry for 400 kV oil-filled cable as examples of sealing ends for paper-insulated cables, and Figure 6.15 shows the three outdoor sealing ends of a 110 kV external gas pressure cable.

Straight joints

Joints for paper-insulated cables, unlike sealing ends, are made entirely on site, although, in this case, paper is used that has been pre-impregnated in the factory. The impregnating medium is again exactly the same as that used for the cables concerned. For oil-filled cables the joints are produced in single-conductor form, with one housing per conductor, while for (three-phase) gas pressure cables they are constructed in a three-phase form in a shared steel-pipe housing.

The on-site installation operation essentially consists of the following steps, all of which must be carried out manually (for further information see Figure 6.16):

- *Making up the ends*: The wires and the conductor are exposed for the lengths required, and a *cone* is produced or the insulation prepared in *steps* at each cable end. The cone, i.e. a tapering of the insulation that is effected as uniformly as possible by pulling off the separate layers of insulation, is preferable for electrical reasons, but does involve more extensive manual work later in the installation procedure.

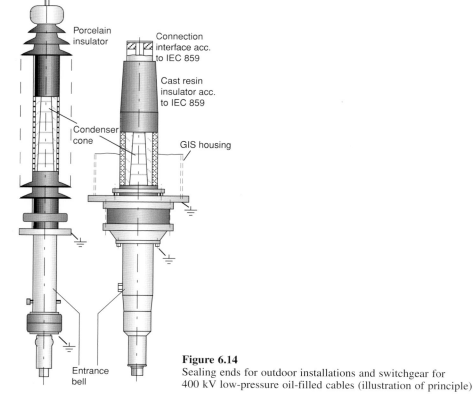

Figure 6.14
Sealing ends for outdoor installations and switchgear for 400 kV low-pressure oil-filled cables (illustration of principle)

Figure 6.15
Outdoor sealing ends of a
110 kV external gas pres-
sure cable

- *Connecting the conductor*: The conductor is connected using one of the techniques
  described in Section 6.2.1. One of the equal-diameter types of connection is generally
  used with extra high-voltage cables for electrical reasons.

- *Connector smoothing*: To compensate for field disturbances around the conductor connection, conductive paper (carbon paper) is wound around; in the case of larger-diameter conductor connections, elastic crepe paper is used.

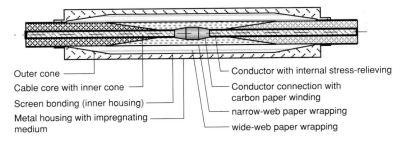

Outer cone
Cable core with inner cone
Screen bonding (inner housing)
Metal housing with impregnating
medium

Conductor with internal stress-relieving
Conductor connection with
carbon paper winding
narrow-web paper wrapping
wide-web paper wrapping

Figure 6.16   Illustration of the steps involved in producing a joint for paper-insulated cables

195

- *Compensation winding*: In the case of larger-diameter conductor connections, lapped insulation must first be applied using so-called *narrow-web papers* consisting of pre-impregnated rolls approx. 20 mm in width. When the cone technique is used for preparing the end, this type of winding should be applied until the original core diameter is reached. In the case of stepped ends the winding should stop when the diameter differences at the point of connection have been evened out.

- *Wide-web insulation*: The majority of the insulation for the joint is produced from pre-impregnated paper webs up to 1.5 m in width, which should be cut in advance to the appropriate size when the end has been prepared in steps. The total thickness of the joint insulation is generally 1.5 to 2 times the cable insulation, because the manually applied dielectric cannot achieve the high stress withstand values offered by the cable insulation that has been manufactured using machines. As a result of this, the heat dissipation conditions in the area surrounding a joint may sometimes be less effective than in the undisturbed cable.

- *Shielding* and *shield bonding*: The insulation is covered with carbon paper or Höchstädter paper at ground potential in order to produce a void-free contact with the joint insulation and to connect the cable sheaths so that they are capable of carrying current, e.g. with Cu wire braid. The insulation thus produced is virtually cylindrical, which is the reason that in the area surrounding a joint – unlike the sealing ends – no capacitor foils are required for potential grading.

- *Joint housing*: Joints in high-voltage paper-insulated cables are generally surrounded by a metal housing (e.g. aluminum), which is provided with suitable protection against corrosion (e.g. bitumen) on the outside.

- *Evacuation and filling*: The installation of the joint, as with sealing ends, is followed by a lengthy process of evacuation and filling with a suitable impregnating medium for the cable insulation concerned (low-viscosity liquid or high-viscosity compound).

All in all, the installation of accessories for high-voltage and extra high-voltage paper-insulated cables requires a considerable degree of know-how and care, and can only be carried out by trained personnel. Operations of this kind have therefore until now – unlike many operations with 110 kV polymer-insulated cable – been carried out exclusively by fitters employed by the cable manufacturers and not by the end user's staff.

Joint for single-phase LPOF cables

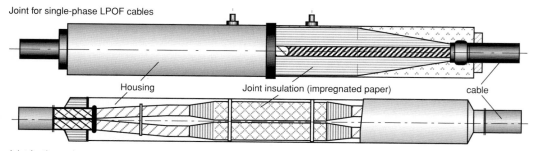

Housing   Joint insulation (impregnated paper)   cable

Joint for three-phase gas pressure pipe type cable

Figure 6.17
Illustration of principle of straight joints for low-pressure oil-filled cables (above) and gas pressure cables in steel pipes (below)

Cross-sectional drawings of single-phase and three-phase straight joints for high-voltage cables with a paper dielectric are shown in Figure 6.17. The oil-filled cable joint also illustrated there is a version with screen separation, i.e. a metal housing divided by an insulating ring as required for the cross-bonding of the cable sheath, as described in Section 3.4.1, Figure 3.22. See Section 6.3.3. for details of the construction of such *separating joints*.

6.2.3 Accessories for Polymer Insulated Cables

The fundamentally simpler structure of the insulation in XLPE cables led to the early development of easy-fit accessories, whose most important electrical components, the *field control elements*, are prefabricated and tested in the factory. Such accessories simply need to be push-fitted to an appropriately prepared cable on site (*slip-on stress cones, slip-on joints*). Many more years passed before this approach, which was initially developed for medium voltage applications, was successfully upgraded for the high-voltage and extra high-voltage fields. There is now consequently a large number of alternative solutions competing with these prefabricated elements, and as a result the range of products available is considerably broader than in the field of accessories for paper-insulated cable. In the mid 1990s the Cigré Working Group 21.06 produced a comprehensive overview of the large number of accessories available [37], and this section sets out to describe only the most important types:

- Slip-on sealing ends and joints

- Sealing ends and joints with a modular design

- Lapped sealing ends and joints

- Extrusion molded joints

Table 6.2 provides an overview of the application areas and nominal voltages for which the techniques described in detail below are suitable.

Slip-on sealing ends and joints

Accessories with slip-on stress cones for high-voltage and extra high-voltage cables generally use the geometric-capacitive method of field control. As shown in Figure 6.7, suitably contoured deflectors made from an elastic conductive compound are bedded void-free into a similarly permanently-elastic insulator and are finally pressed in one piece

Table 6.2 Application areas of the various types of accessories for polymer-insulated cables

Type	Dielectric	Accessory	≤150 kV	≤275 kV	≥400 kV
Lapped	PP, PPL + control layers	SE	◆	◆	◆
Lapped	Self-amalgamating tapes	Joint	◆	–	–
Components	SIR, EPR, cast resin	SE, joint	◆	◆	◆
Slip-on, one-piece	SIR, EPR	SE, joint	◆	◆	◆
Extruded	XLPE	Joint	–	◆	◆

onto the appropriately prepared polymer-insulated cable (LDPE, XLPE or EPR) and positioned precisely.

- *Slip-on stress cones*

Control elements for sealing ends generally have only one deflector, which is fitted over the stripped edge of the outer conductive layer and effectively reduces field magnification at this point. To avoid discharges on the surface of the control element, it is placed in an external insulator, which is generally filled with an insulating fluid (high-viscosity polyisobutylene PIB or silicone fluid, in special cases even low-viscosity ester fluids [132]) or maintained under increased gas pressure (SF_6) [133]. Using this kind of embedding medium, however, does not conform to the concept of a completely maintenance-free system, which is the goal that engineers are striving to attain in connection with polymer-insulated cable systems. The introduction of so-called *dry sealing ends* could help in this respect.

The shape and materials of the external insulator are determined by the application, as with paper-insulated cables: porcelain or composite materials as shown in Figure 6.13

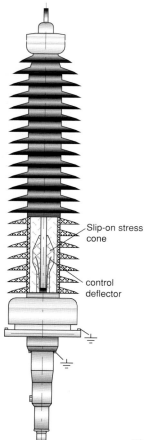

Figure 6.18 Outdoor sealing end with slip-on stress cone (schematic)

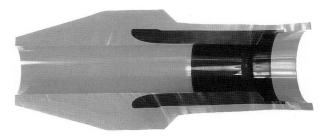

Figure 6.19   Sectional view of a slip-on stress cone (black part: conductive deflector)

for outdoor use and cast resin for transformer and switchgear entries. As an example of the use of prefabricated slip-on stress cones, Figure 6.18 shows the cross section of a high-voltage cable sealing end for outdoor use; the construction of the associated control element with its conductive deflector is shown by the cutaway photograph of a slip-on stress cone in Figure 6.19. At higher voltages this simple design can only be implemented with slip-on stress cones made from silicone rubber or EPDM which do not require any external pressurization.

- *slip-on joints*

  Modern developments are increasingly moving towards prefabricated and factory-tested slip-on units even for straight joints for polymer-insulated cables. All components needed for field control are already built into these units [80, 134]. As illustrated in Figure 6.4, in a greatly simplified form the requirement for field control in joints can be concentrated on the conditions of two opposing cable ends with the conductor connection between them. The slip-on unit consequently needs two opposing control deflectors, not unlike those that were described above for sealing ends, and a field-smoothing covering for the conductor connection. Finally, a conductive coating of the surface of the joint is needed to provide the outer screening.

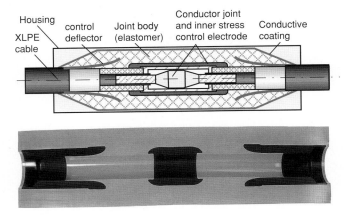

Figure 6.20
Cross section of a prefabricated slip-on joint for polymer-insulated cables, schematic drawing and photograph

199

Figure 6.20 shows the cross section of this type of joint in the form of a schematic drawing as well as a photograph. To protect against the ingress of moisture and against mechanical damage the slip-on joint is installed in a metal housing that is protected appropriately against corrosion and can then be buried directly in the soil.

Despite the use of deflectors with the same structure (*modular technology*) which, not least, has the positive effect of reducing the cost of forming in the manufacturing process, different field conditions are found in joints than in sealing ends. In particular, the *tangential component* of the electrical field that is locally strengthened at the interface between cable dielectric and joint insulation becomes more significant relative to the maximum field strength within the body of the joint. In order to avoid unacceptably high stresses and to optimize the shape of the joint, separate field calculations, as described in Section 6.1.2, are consequently necessary. Figure 6.21 shows the results of corresponding calculations in the form of the potential distribution and the field pattern for the normal and tangential components in a 400 kV slip-on joint, as in Figure 6.20.

The thickness of the joint insulation significantly exceeds the wall thickness of the cable insulation in the area around the junction. This results in the possibility of local overheating even in prefabricated *slip-on joints*, which must be counteracted in an appropriate way.

Slip-on technology require in particular a high degree of permanent elasticity alongside good electrical and thermal characteristics in the materials involved. These properties are necessary to ensure that the contact force of the control elements on the surface of the conductor is sufficient under all circumstances despite the thermal load cycles and accompanying contraction/expansion of the polymer-insulated cables. Only two materials meet these criteria: *Silicone rubber* (SIR) and synthetic rubber (EPR or EPDM). Sil-

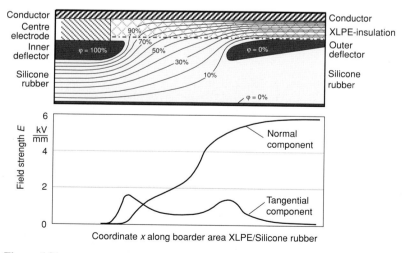

Figure 6.21
Calculated potential distribution and field patterns in a prefabricated SIR slip-on joint for 400 kV XLPE cables

icone rubber is the more flexible of the two, and, since as a two-component system, it is self-hardening. It is also easier to work with than EPR; on the other hand EPR is more mechanically resistant, less sensitive to the effect of liquid embedding media in sealing ends and considerably less expensive than SIR, although it does require very costly pressure-resistant vulcanizing molds. For slip-on stress cones for XLPE extra high-voltage cables ≥ 400 kV, both silicone rubber and EPR/EPDM have meanwhile become widely used [135].

Until recently, it was virtually unthinkable from both the manufacturing and the installation point of view to use one-piece slip-on units – at least for joints in 400 / 500 kV systems – because of their dimensions. Thus, for example, not one single manufacturer of the six who were involved in the Bewag pre-qualification test for 400 kV XLPE cable systems concluded in 1995 [129] has actually used this type of slip-on joint. Further, in two reports published by Cigré in 1996, any major chances for the so-called *component joint*, which is assembled on site from several components, were considered to be poor [136, 137]. Costly testing and consequent improvements in materials [138], in conjunction with a reduction in the size of field control elements by using modified deflector contours [131, 139] based on the numerical field calculation (Figure 6.21), proved the only way of producing one-piece slip-on joints that could be fully electrically tested in the factory. The installation techniques were also perfected by designing suitable electro-hydraulic push-fitting tools that enabled even large joint bodies to be positioned successfully with millimeter precision and without subjecting the elastomer to any excessive mechanical strain. Development has meanwhile progressed to the point where a 6.3 circuit-km system to extend the inner-city 400 kV network has been implemented in Berlin with XLPE cables using this type of joint technology.

The *installation* of a slip-on joint on XLPE cables comprises a total of seven steps, which are shown in schematic form in Figure 6.22

1. The sheath is stripped back as far as the outer conductive layer to different lengths on the two cable ends as the body of the joint must first be put in a "park position" at full length on one side before the conductor connection is produced.

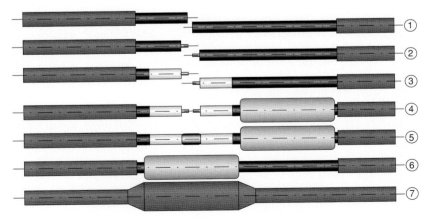

Figure 6.22   Assembly of a slip-on joint (see main text for explanation)

2. The conductors are exposed by cutting back the cable insulation in a straight line; there is no need for a conical shape as with lapped joints.

3. The outer conductive layers are removed over the same length from both cable ends, and the stripped edges and core surfaces are carefully prepared in the area in which they come in contact with the joint insulation.

4. The body of the joint is pushed on one cable end to the specified "park position".

5. The conductors are connected.

6. The body of the joint is positioned centrally over the conductor connection using electro-hydraulic pushing devices and a compatible lubricant.

7. The outer conductive layers of both ends of the cable are connected with the conductive coating of the joint body[1] and the housing is assembled.

Preparation of the core surface requires great care. The most critical factor is the quality of the stripped edge and of the contact area between cable insulation and joint insulation, which ultimately determines the long-term electrical behavior of the interface. This is influenced by a number of factors:

- *Surface roughness* of the core insulation and joint insulation,

- Type and quantity of lubricant used,

- Care taken and precision achieved in the process of pushing the joint on and the final positioning of the joint body,

- Contact force and material characteristics of the body of the slip-on joint (elasticity, long-term mechanical behavior).

Even when the installation operations are performed with the utmost care and the mechanical and electrical design of the joint is perfect, the operational stress on the outer conductive layer of the joint should not exceed 6 to 7 kV/mm. This should be taken into consideration when dimensioning the thickness of the insulation walls of the cable and joint body.

Modular design for sealing ends and joints

Accessories of modular design represent in principle a precursor of the one-piece slip-on units: Again, the central aims of accessory design are to be able to largely prefabricate and test the main components in the factory and to install it in the most straightforward manner possible on site. The forefront of developments in this area was in Japan, initially for extra high-voltage sealing ends, and the expertise acquired was transferred later to joints [136, 140, 141]. In order to avoid possible problems in manufacturing and installing extremely large single-piece control elements, they have been broken down into various smaller components which then need to be re-assembled on the construction site:

---

[1] For details of the implementation of screen separation in separating joints see Section 6.3.3.

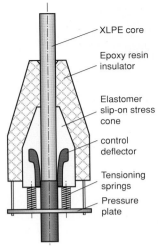

Figure 6.23
Schematic drawing of the elements of a modular-design cable sealing end that are used for field control

- Modular-design *sealing ends* have at least three components, as shown in Figure 6.23. These are an epoxy-resin insulator, a relatively small slip-on element for inserting in this insulator which includes a geometrically / capacitively operating control deflector made from elastomer material (generally EPR) and a tensioning mechanism with helical springs which provides the necessary compression for the elastomer component between the surface of the core and the insulator. Sealing ends of this type are now available for XLPE cables up to 500 kV [142].

- The plug-in sealing ends for switchgear and transformers up to max. 170 kV, described in Section 6.2.1 (Figure 6.12), should also be included in the category of "modular-design accessories" since they are assembled from several component parts on site. Nevertheless, they are distinguished from other sealing ends designed for the same purpose in that the switchgear or transformer transition housing does not need to be opened in order to install them – as long as the latter were originally equipped with the appropriate coupling elements as shown in Figure 6.12.

- The design of *modular-design joints* is derived from that of sealing ends, and they need five separate components for field control (Figure 6.24): the cast-resin insulator with integral high-voltage electrode in contact with the conductor connection, one slip-on element with control deflector on each side of the insulator, and the two tensioning

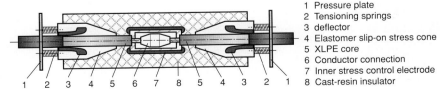

1 Pressure plate
2 Tensioning springs
3 deflector
4 Elastomer slip-on stress cone
5 XLPE core
6 Conductor connection
7 Inner stress control electrode
8 Cast-resin insulator

Figure 6.24
Conceptual drawing of the elements of a modular-design straight joint that are used for field control

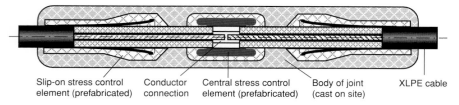

Slip-on stress control   Conductor   Central stress control   Body of joint   XLPE cable
element (prefabricated)   connection   element (prefabricated)   (cast on site)

Figure 6.25   Silicone rubber component joint as described in [137]

devices with springs for pressure application that are needed to compress the slip-on elements between the insulator and the conductor surface.

- Unlike the accessories described here, the modular-design joint presented in [137] has no separate cast-resin insulator or slip-on stress cone on each side. Instead, the remaining joint insulation is cast on site in cold-hardening SIR using a disposable cast. This operation is performed after the prefabricated, tested slip-on SIR joint, along with a likewise prefabricated slip-on unit that covers the conductor connection, have been assembled at a specified pre-tension as shown in Figure 6.25. A certain reduction in volume occurs during the vulcanization process, and this builds up a long-lasting contact pressure on the cable core. The end result of this operation is a single-piece joint body made from a single base material (SIR) which is virtually indistinguishable from the "truly" single-piece slip-on joint illustrated in Figure 6.20.

- The so-called *back-to-back joints* fall in between the other multi-piece accessories; they are, in fact, nothing more than two switchgear sealing ends joined together (Figure 6.26). This technology has until now been used in France at 225 kV but, more importantly, in an extensive PE cable system at 400 kV. It offers the advantage that existing components can be used and removes the need to develop separate joint units or to maintain the associated manufacturing equipment [142].

As well as the back-to-back coupling of two complete sealing ends with two cast-resin insulators, it is also possible to reduce material costs by using only one insulator, or even to directly connect two slip-on sealing end stress cones back to back in a shared section of pipe with no insulator at all, but with SF_6 at increased pressure. Irrespective of such configuration details, however, back-to-back joints are left with two crucial disadvantages that stand in the way of their more widespread use:

- The extremely large dimensions of a back-to-back joint and, most importantly

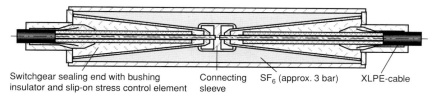

Switchgear sealing end with bushing   Connecting SF_6 (approx. 3 bar)   XLPE-cable
insulator and slip-on stress control element   sleeve

Figure 6.26
Principle of the "back-to-back joint" comprising two opposingly coupled switchgear sealing ends

- The necessity of supplying SF$_6$ at increased pressure, and the consequent need for pressure monitoring, which cancels out one of the main advantages of polymer-insulated cables, namely the low level of maintenance required.

Lapped sealing ends and joints

One drawback shared by all prefabricated units lies in the need to match each unit individually and precisely to the cable dimensions involved, which makes it necessary to stock an extensive range of suitable forming tools. Lapped accessories, on the other hand, can be matched to a great range of conductor cross sections and insulating wall thicknesses without any problem.

The fundamental principle of lapped accessories was transferred directly from paper-insulated cable technology to polymer-insulated cables, with alternative materials having to be found in place of the paper tapes and webs. The only suitable materials for this purpose are those exhibiting sufficient permanent elasticity to be able to respond to the considerable changes in volume that are typical of polymer-insulated cables under fluctuating thermal loads [140]. Depending on the cable dielectric (LDPE, XLPE, EPR) and the nominal voltage, the following are the main materials used for lapped accessories:

- *Polypropylene foils* for sealing ends with capacitor field control, whereby the required conductivity of the control layers is achieved by adding extra carbon black or graphite. This is similar to the technique used with the semi-conductive compounds for the field-smoothing layers in the cable. In order to avoid premature discharges, the capacitor cone is impregnated with silicone fluid, polybutene, or SF$_6$ at increased pressure. Depending on the application, either porcelain or GRP insulators are used to provide external protection. Any differences between the construction of these sealing ends and that of lapped sealing ends with paper insulation (as shown in Figure 6.14) are thus so slight as to be unnoticeable. This technology has been used in Japan with XLPE cables up to 500 kV [79].

- *Polyethylene or cross-linkable PE tapes* as joint insulation for thermoplastic PE cables and XLPE cables. So-called *spliced joints* with this type of dielectric offer the advantage of very compact dimensions, and amongst other applications they have been used in Germany with the thermoplastic PE cables installed initially for 110 kV, and were later also developed and tested for XLPE cables for voltages up to 400 kV. A number of problems were encountered, however: points where bonding had been performed less than perfectly brought the danger of partial discharges [143]. Equally problematic were the extreme requirements for cleanliness in on-site assembly along with the complicated regulation of temperature and pressure for curing, and above all the cross-linking of the winding tapes applied manually or semi-automatically [143]. This approach was therefore largely superseded again by other solutions.

- *Self-amalgamating EPR tapes* as a material for joints for XLPE and EPR cables, which are produced largely automatically using *lapping machines* as shown in Figure 6.27. The amalgamating of the tapes is triggered by specified over-stretching during the lapping operation, with the result that these joints can be produced without using any increased temperatures [37, 143]. Unlike polyethylene, moreover, EPR insulation tapes show little tendency to develop electrostatic charge because of their higher level of basic conductivity; the danger of dust accumulating or even sparks developing as a result of an intrinsic electrical field is therefore so slight as to be negligible. EPR also has a significantly lower susceptibility to PD than PE/XLPE.

Figure 6.27   Lapping machine for producing joints from self-amalgamating EPR tapes

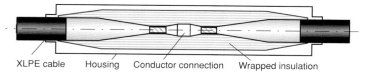

XLPE cable   Housing   Conductor connection   Wrapped insulation

Figure 6.28   Straight joint formed from self-amalgamating EPR tapes (schematic)

The greater intrinsic conductivity of EPR, on the other hand, is responsible for the relatively high dielectric losses in the insulation. This in turn restricts the use of this technology to high-voltage cables up to 150 kV even though the technology is generally unproblematic and has been widely tried and tested. The basic construction of an EPR lapped joint is illustrated in Figure 6.28.

Extrusion molded joints

Extrusion molded joints (EMJ) are based on the idea that optimum accessory insulation should be structured in exactly the same way as the cable insulation itself. In the case of XLPE cables, this means that the joint dielectric in the form of cross-linkable polyethylene is extruded on-site onto the prepared conductor connection, which has been spread with a semi-conductive layer of XLPE. An external form (mold) is used for this purpose, and the dielectric is subsequently cross-linked thermally under external pressure [144]. Figure 6.29 shows the fundamental structure of the associated manufacturing equipment which needs to be installed in clean-room conditions on site in order to avoid impurities in the joint dielectric that would reduce its strength.

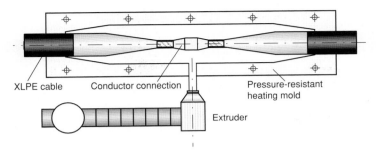

Figure 6.29   Equipment used in the production of extrusion molded joints

This technology, which was first developed and applied in Japan for 275 kV cables, has meanwhile been successfully transferred to the 500 kV field [145, 146], and one of the factors that sets it above the alternatives is its particularly slim construction (insulation thickness for 500 kV only 36 mm [10]). However, not only does the extrusion method involve high on-site costs in terms of the amount of technical equipment required, but the time needed to install the equipment is unusually long, taking a team of highly-qualified engineers up to 30 days. The fact that no manufacturer outside Japan has been using extrusion molded joints can therefore be attributed partly to economic reasons. These reasons also account for the fact that, even in Japan, intensive work is meanwhile being carried out on developing alternatives based on prefabricated components [23].

6.3 Special Versions of Joints

The "special versions" of joints described in this section are those that, in addition to their standard task of linking conductor with conductor and screen with screen (or sheath with sheath), also perform certain extra functions. There are three special versions:

- Stop joints for oil-filled cables

- Transition joints for XLPE / paper-insulated cables

- Separating joints (cross-bonding technology)

6.3.1 Stop joints for oil-filled cables

The function of stop joints for low-pressure oil-filled cables, as described in Section 4.15, Formula 4.9 and Figure 4.17, is to hold the pressure of the insulating oil within specified limits (approx. 2 bar $\leq p < 8$ bar), e.g. in sections where there is a considerable height drop. This is done by dividing up the whole run into several hydrostatically separate shorter sections.

In a certain sense, the construction of stop joints is similar to that of back-to-back joints for XLPE extra high-voltage cables (Figure 6.26). As Figure 6.30 clarifies, stop joints similarly consist of two cable sealing ends linked back to back in a common grounded

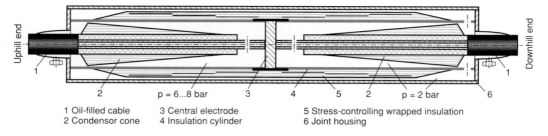

1 Oil-filled cable   3 Central electrode   5 Stress-controlling wrapped insulation
2 Condensor cone   4 Insulation cylinder   6 Joint housing

Figure 6.30   Design of a stop joint for oil-filled cables (schematic)

joint housing. The dielectrics and field control systems used are suitable for the technology generally used with accessories for paper-insulated cables, i.e.:

- Both sealing end insulators contain capacitor cones made from impregnated paper, and

- The insulation on the conductor connection opposite the common joint housing is also paper (as opposed to the SF_6 used in the back-to-back joint). A special system of capacitive layers for field control is required in addition between the conductor connection located at the high-voltage potential and the grounded housing. Both the bushing insulators deviate from their usual shape and have a cylindrical external contour for accommodating the controlled paper insulation. The low-viscosity impregnating medium in this insulation is in contact with the impregnating medium of the *lower* insertion insulator; the system pressure is approx. 2 bar (the same as the pressure at the *upper* end of the *deeper* section of cable). Separated from this is the fluid system of the other (upper) bushing insulator; it is in contact with the higher section of cable, and is under a pressure of max. 6 to 8 bar.

In place of the controlled lapped paper between the conductor connection and the grounded joint housing, an epoxy-resin insulator with a fixed, cast high-voltage electrode may be provided. Such an insulator simultaneously handles the hydrostatic separation between the oil reservoirs on each side.

Stop joints of this kind are therefore extremely complicated and equally costly, but they contain three complete, integrated and largely controlled insulation systems between high-voltage and ground potential. Wherever possible, this type of component is not used today, and engineers are falling back on other types of cable instead (gas pressure cables or XLPE cables depending on the voltage level and transmission capacity, or – particularly for a short link to an underground power plant – also SF_6-insulated tubular conductors as shown in Figure 5.16).

6.3.2 Transition Joints between XLPE / Paper-Insulated Cables

If XLPE cables are to be connected to existing paper-insulated cable systems in the course of network extensions or replacement of damaged sections, then suitable straight joints must be used. One side of these must be compatible with the "paper" insulation system, and the other with XLPE. This function is performed by *transition joints*, for which – to some extent depending on the type of paper-insulated cable that is to be con-

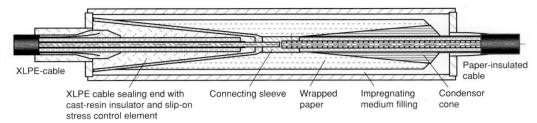

XLPE-cable

XLPE cable sealing end with
cast-resin insulator and slip-on
stress control element

Connecting sleeve

Wrapped
paper

Impregnating
medium filling

Condensor
cone

Paper-insulated
cable

Figure 6.31
Construction principle of a single-phase transition joint between XLPE and paper-insulated cables

nected with the XLPE side – extremely diverse solutions have been developed. These can be divided into three main categories [37]:

- *Three-phase transition joint*

- *Single-phase transition joint* between XLPE and paper-insulated cables with *no* hollow channel in the conductor

- *Single-phase transition joint* between XLPE and paper-insulated cables *with* hollow channel in the conductor

The first two categories are intended mainly for gas pressure cables, while the third version is used for connecting XLPE with low-pressure oil-filled cables.

Just one example will be described here that lends itself, with only slight modification, to both gas pressure cable and oil-filled cable. This transition joint also exhibits similarities with the back-to-back construction of two sealing ends connected in opposition (Figure 6.31).

On the XLPE side, a conventional slip-on stress control element with a geometric-capacitively acting deflector, as illustrated in Figures 6.18 and 6.19, takes care of the field control; on the paper-insulated cable side, prefabricated capacitor cones with conductive control layers (Figure 6.5) are used. The separation between the XLPE side and the paper-insulated cable side, which is under increased pressure, is ensured by the cast-resin insulator of the XLPE cable sealing end, which in this case is cone-shaped. This insulator is filled with polyisobutylene (PIB) or another XLPE- and SIR-compatible insulating liquid in order to avoid partial discharges on the surface of the slip-on stress cone [147].

The conductor connection between the two cable sealing ends is insulated from the grounded joint housing – as with the stop joint for oil-filled cable – by wrapped paper that is fitted on site. In the case described here, this insulation is sufficient without any additional control layers (while alternative designs even provide fixed cast-resin insulators with an embedded control electrode at conductor potential [37]). The joint housing is filled with the impregnating medium from the paper-insulated cable side (low-viscosity oil or high-viscosity compound), and is also under the system pressure of that side (i.e. 2 to 8 bar with low-pressure oil or 15 bar with gas pressure cables). In the case of low-pressure oil-filled cables, a transverse hole through the tubular conductor ultimately ensures that an exchange of impregnating medium takes place between the joint insulation and the cable insulation.

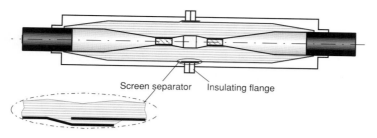

Figure 6.32
Separation of the outer, field-smoothing layer in cross-bonding joints with lapped insulation

6.3.3 Separating Joints

These joints, otherwise known as *transposition joints* or *cross-bonding joints*, have the task of ensuring that the cable sheaths or screens on the two sides are electrically isolated and sufficiently surge-proof (Section 8.2) so that the current losses from the sheath can be reduced by application of the *cross-bonding* technique as described in Section 3.4.1 (Figure 3.22). In order to perform this function, separating joints differ from conventional versions in the following design features:

- Screen or sheath connections on both sides of the cable that are accessible from outside and are of sufficient size to carry the induction current occurring, even under short-circuit conditions,

- Separator in the outer, field-smoothing layer on the joint surface and

- Separator in the metal joint housing.

While the joint housing is most easily provided with a separator by installing a sufficiently wide cast-resin insulating flange, the positioning and type of separation for the conductive layer must be tailored in a suitable way according to the structure of the joint insulation. "Suitable" in this context means that the separation point must in no way adversely affect the virtually cylindrically symmetrical field distribution within the joint insulation. This is achieved

- In lapped joints by overlapping the outer conductive layer with a thin intermediate layer of insulating material as shown in Figure 6.32 and

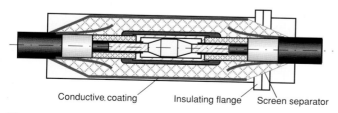

Figure 6.33
Separation of the outer, field-smoothing layer in cross-bonding joints in either the slip-on or modular-design version for polymer-insulated cables

- In prefabricated joint bodies (one-piece or modular design) for polymer-insulated cable by discontinuing the outer conductive layer in the weak-fielded or virtually field-free area behind one of the control deflectors on the cable ends (Figure 6.33).

In both of the above cases the insulating flange of the joint housing is positioned precisely over the separation point in the conductive layer. Nonetheless, it is impossible to avoid completely a slight field disruption in the area surrounding the separation points, and that is why – especially in the case of taped joints with paper or plastic dielectric that need to be produced on site – electrical isolation should only be used if there is no alternative.

7 Cable Systems

The term "cable system" covers all the components necessary for the underground transmission of electrical energy. In addition to the cables themselves and their accessories, this also includes various measuring and protective devices, along with equipment for maintaining and monitoring pressure, monitoring liquid impregnating media, embedding media or compressed gases that may be used according to the cable type. In recent years these have been joined by systems for continuous precautionary observation of particular cable properties, for which the term '*monitoring*' has become the established norm. After being produced and tested by the manufacturer the components are installed on site by trained engineers, then tested again before finally being commissioned. These on-site installation operations, which in general account for well over 50% of the total investment cost for the cable system, essentially comprise two activities: *Laying* the cables and *installing* the accessories and additional components.

7.1 Laying and Installation

7.1.1 Route Planning

The first step when installing a new cable system is to define the route. In order to minimize the cost it is desirable to link the start and end points of the cable system by the shortest possible route. However, in the usual environment for high-voltage or extra high-voltage cable runs – i.e. in cities or other densely populated areas – compromises generally need to be made in the pursuit of this goal since the shortest possible link (a straight line from point A to point B) is usually impossible because of structural obstacles. Another factor that generally excludes the shortest route when planning a cable connection is the legal aspect in terms of property ownership; planners therefore favor the courses of public roads and paths, which very rarely enable the link to follow a straight line.

After the general outline of the route has been established, the next stage is to *plan it in detail*. This stage involves planning the precise course of the cable trench, taking into account the minimum bend radius for high-voltage cables with polymer, lead or corrugated aluminum sheaths as shown in Table 7.1, and also the location of the straight joints that may be necessary between the separate sections supplied. Here, too, consideration must be given to more than one aspect:

- The lengths supplied should be as long as possible, and the joints required correspondingly as few as possible.

- There must be sufficient space at the junction points to allow access for heavy machinery, and

- In the event that *cross bonding* is to be applied to reduce sheath losses, the sections between the joints must, if at all possible, be of the same length and their number must always be an integer multiple of three (Section 3.5.2).

Finally, in the context of route planning, the *laying depth* of the cables must be specified. High-voltage and extra high-voltage cables must always be routed below all other existing or planned installations, and they must be protected against accidental damage either mechanically with steel or concrete plates, or at least visually with suitable warning tape. The minimum laying depth for high-voltage cables below public footpaths in Germany is 1.2 m, and below roads it is even greater at 1.35 m. Factors concerned with protecting the cables from vibrations must also be considered here, and guidelines have also recently been issued on limiting the magnetic field developed by such cables at ground level in public places (see Section 7.3).

In practice, new high-voltage cable systems in centers of population will generally be buried considerably deeper than 1.2 to 1.35 m, since levels closer to the surface are already occupied by power cables of lower voltage ranges and telephone cables, along with supply lines for water, gas, district heating and various other services. It is not uncommon for cables to be buried at a depth of 2.5 to 3 m. On the one hand, large distances from the surface like this mean – in line with the equivalent thermal diagram of underground cables in Section 3.5.1 (Figure 3.30) – higher thermal resistances of soil and a resultingly lower load capability of the cable system; on the other hand the layers of soil in question are warmed considerably less in the summer than the soil close to the surface. Figure 7.1 gives an idea of the seasonal fluctuations in central European soil temperature at various depths. The smaller temperature rise in summer at greater depths below the surface lends naturally-cooled cables an increased current carrying capacity with the net result that the influence of the higher thermal resistance of soil is canceled out almost completely.

Rather than laying the cables in open trenches, systems and, in particular, extra high-voltage systems, are sometimes routed in tunnels, one method of forming the tunnels being the so-called *shield driving method*. This offers a way of avoiding the difficulties associated with long-term large-scale open building sites and the unacceptable traffic problems they cause. This method was used, for example, in implementing the first 400 kV XLPE cable run for extending the inner city extra high-voltage network in Berlin [105, 127].

Gas pressure cables in steel pipes are a special case and their route is often constructed long before the actual laying of the cables. When excavation operations are taking place anyway, this can be taken as an opportunity to install initially empty steel pipes underground without adding significantly to the length of time taken by the works. Cables can

Table 7.1
Minimum permissible bend radii R_{min} for high voltage and extra high voltage cables

Insulation	No. of cores	U_N in kV	R_{min}
Paper	single-phase	≤110 >110	$25 \times D$ $35 \times D$
Paper	three-phase	≤110 >110	$15 \times D$ $25 \times D$
Polymer	single-phase	60…500	$15 \times D$

D = Diameter over outer sheath

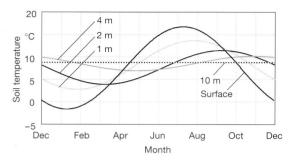

Figure 7.1 Seasonal fluctuations of soil temperature in central Europe

then be drawn into these pipes later when required, and operational pressure applied subsequently. The only maintenance needed for the initially empty pipe system is protect the inside of the pipe effectively against damp to enable the cable dielectric to be introduced at the appropriate time to an absolutely dry atmosphere. For this purpose, the pipes are usually filled with dry nitrogen and maintained at a small positive pressure.

The individual lengths of cable needed can only be determined and ordered from the cable manufacturer once the type of cable run is known as well as its precise course and the position and size of the joint structures (*joint trenches* that are filled in with backfill material after the joints have been installed, or *joint chambers* that remain permanently accessible). Further relevant factors are the laying depth and – in the case of low-pressure oil-filled cables with a significant gradient – also the positions of any stop joints that may be required (see Section 6.3.1).

7.1.2 Choice of System Configuration

The term *system configuration* refers to the arrangement of the three phases of a cable system relative to one another. The main distinction is between a *flat formation* and a *trefoil layout*; with both of these systems, the distance between the axes of the phases, the type of cable sheath and, above all, the grounding conditions are of the utmost importance. The system configuration has an effect both on the current-dependent losses of cables through the proximity effect and on sheath voltage induction (Section 3.4.1), as well as on the so-called *electromagnetic interference* with other underground lines and possibly on people or animals in the vicinity of the cable system (Section 7.3). To some extent the heat dissipation from the cable system also depends on its configuration.

The effects operate to a certain extent in opposition to one another, i.e. a configuration that has a positive effect on the sheath losses or electromagnetic interference can prove to be unfavorable in terms of current displacement and/or heat dissipation, and vice versa. For clarification purposes, Table 7.2 shows the effects of different system configurations on the following four operational characteristics of a cable system:

– Conductor losses through the proximity effect,
– Sheath losses through induction currents,
– Heat dissipation into the surrounding soil and
– Electromagnetic interference of the surrounding environment.

Table 7.2 The effect of system configuration on various operational characteristics

Configuration change	Effect on proximity effect	Effect on sheath losses	Effect on heat dissipation	Effect on interference
⊙⊙ instead of ⊙⊙⊙	unfavorable	favorable	unfavorable	favorable
⊙⊙ instead of ⊙ ⊙ ⊙	favorable	unfavorable	favorable	favorable
⊙⊙⊙ instead of ⊙⊙⊙	none	none	indifferent	favorable[1]
▭ instead of ▭	none	unfavorable	none	favorable
Cross bonding instead of ▭	none	favorable	none	none
Al sheath instead of lead	none	unfavorable	none	favorable

[1] Applies only to objects at surface level, otherwise no effect

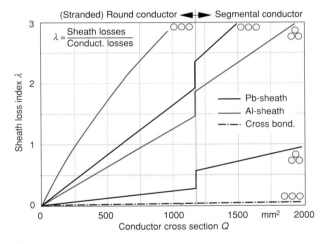

Figure 7.2
Relative sheath losses with various different system configurations and sheath grounding on both ends

It shows that cross bonding represents the only intervention in the configuration, which – apart from considerably increased investment costs – does not appear to have any adverse side effects. This is also borne out by Figure 7.2, which supplements Table 7.2 by showing the quantative pattern of sheath losses in a cable system over the conductor cross section with various system configurations. Section 7.3 provides detailed information on electromagnetic interference.

7.1.3 Pulling in the Cables and Installation of Accessories

The laying of high-voltage and extra high-voltage cables and the installation of accessories are – at least in Germany – generally performed by the relevant specialist departments of the companies manufacturing the cables. The normal practice is different in the low- and medium voltage fields, where cable laying and accessory installation are mainly performed by the organizations operating the systems. This is because the economic cost involved in the training and permanent retention of highly-qualified personnel for the relatively rare occasions on which a high-voltage cable system is newly installed is simply too high for the individual operator. On the other hand, certain large power supply companies are, at present, reorganizing themselves to develop their own repair divisions to deal with damage to cable systems, particularly for the relatively easy-to-install polymer-insulated cable systems.

The force required to draw high-voltage and extra high-voltage cables in off the transport drums (which are usually delivered on suitably-equipped *cable trailers*, as illustrated in Figure 7.3) into prepared trenches, tunnels, cooling pipes or steel pipes increases in proportion to the length of cable and its weight per metre. Detours, bends and uphill gradients in the routing are further important factors. In order to minimize the tractive forces required, trenches and tunnel sections are equipped with easy-run *cable rollers* every 3 to 5 m as illustrated in Figure 7.4, which prevent the sheath rubbing on the subsoil. The same function is performed by vertical *corner rollers* positioned on bends in the route. The cable temperature must also not be allowed to fall below certain minimum values during the laying operation in order for it to maintain sufficient flexibility. For paper-insulated cables these *laying temperatures* are at least $+5\,^{\circ}\mathrm{C}$, and for XLPE at least $-5\,^{\circ}\mathrm{C}$.

Assuming that all the necessary rollers are provided, a largely level route and appropriate temperatures, the tractive force required to lay the cables F_Z can be calculated approximately in N:

$$F_Z = f_T \cdot m' \cdot l \tag{7.1}$$

Figure 7.3
Pulling in a 400 kV oil-filled cable from a cable trailer

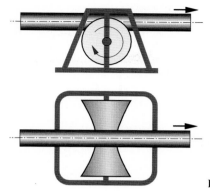

Figure 7.4   Cable roller (schematic)

where

l = Cable length in m

m' = Cable weight per m in kg/m

f_T = *Routing factor* in N/kg in accordance with Table 7.3.

To make it easier to draw cables into existing steel pipes they are armored with so-called *skid wires* made from hard copper or steel. In this case the tractive force required can again be estimated using Formula 7.1, an average friction coefficient of 3 N/kg being applied to arrive at the route factor as in Table 7.3 [148].

Three different methods are commonly used to transfer these forces[1] to the cable [149]:

• *Pulling the cable with motorized rollers*

  Pulling with motorized rollers or caterpillars (Figure 7.5). In the simplest case a specific number of cable rollers, which are needed in any case, are provided along the route with a controllable electric drive and a pressure roller opposite it; the force is applied to the cable sheath. The advantage with this method is that the number of points at which the force is taken up increases in proportion to the length to be pulled off (and consequently increasing transport weight and tractive force). The tractive force to be transferred per force take-up point thus remains more or less constant. Set against this is the drawback of the very costly control technology, which has the task

Table 7.3
Routing factor f_T for calculating the effect of the routing on the tractive power required for laying the cable (see [148])

Route	f_T in N/kg
Trenches, straight	1…4
Trenches with 2 turns, each 90°	2…4
Trenches with 3 turns, each 90°	4…6
Steel pipes (gas pressure cables)	3

---

[1] Example: 400 kV XLPE cable with 1600 mm² Cu conductor, weight approx. 25 kg/m, length $l = 500$ m, Routing factor $f_T = 4$ gives $F_Z = 50$ kN.

217

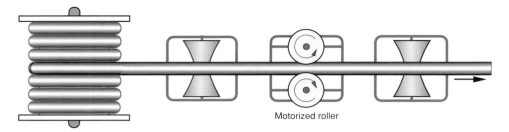

Figure 7.5   Pulling a cable using motorized rollers (principle)

of ensuring that all the drive rollers turn at exactly the same speed and with exactly the same torque whatever the operating status. The result otherwise would be that the cable would bulge out between the drive rollers and this could lead to damage to the sheath through rubbing on the walls or floor of the trench.

- *Pulling a cable in with a rope winch and cable stocking*

  Pulling with the assistance of a rope winch using a *cable stocking* (Figure 7.6). The cable stocking, which is tailored to fit the appropriate cable diameter, is pushed to its full length (≥ 1 m) over the leading end of the cable and secured with wound wire. The whole tractive force must be taken up by the sheath at the leading end of the cable. This method is therefore more suitable for use with cables with robust flat wire armoring (gas pressure cables). The rope winch needed is usually in a fixed installation on a low loader, which is transported by road truck direct to the site.

- *Pulling a cable in using a rope winch and pulling head on the conductor*

  Pulling with the assistance of a rope winch using a *pulling head* on the conductor (Figure 7.7). As with the cable stocking, the whole tractive force is fed in via the leading end of the cable; however, it is applied to the conductor, which is generally considerably more resistant to traction than is the cable sheath. Nominal values are 50 N/mm^2 for copper and 30 N/mm^2 for aluminum conductors. This method is suitable for all high-voltage and extra high-voltage cables, and thus represents the most common solution. Only submarine cables with extreme unit lengths require additional force transmission points in the form of pulling eyes arranged axially on the tension-resistant armor that such cables generally have [148].

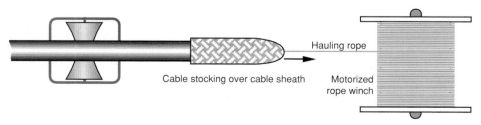

Figure 7.6   Pulling a cable using a rope winch and cable stocking (principle)

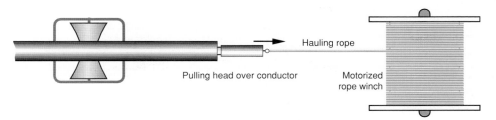

Figure 7.7 Pulling a cable using a rope winch and pulling head on the conductor (principle)

After the cables have been pulled in, installation of accessories commences. Whilst the cable ends are generally accessible, sufficient space for installing the joints must be planned in and cleared when designing the route. Irrespective of the type of cable and the accessories provided for it, the requirement for cleanliness in the installation area is of top priority, since particles of dust and dirt in the already geometrically disturbed electrical field of the accessories can lead locally to a further increase in stress, possibly to partial discharges and eventually to premature failure of the system. While this requirement is usually satisfied in 110 kV installations by an assembly tent, as the rated voltage becomes greater, so too does the cost of ensuring a clean installation environment. With voltages of ≥400 kV it is not uncommon to find dust-proof inner tents being used with filtered air at a slight positive pressure, right through to complete air conditioning for maintaining clean air and a constant ambient temperature. The

Figure 7.8 Outdoor sealing ends of a 400 kV oil-filled cable

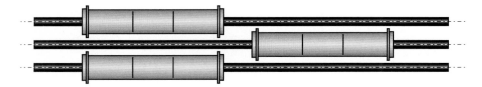

Figure 7.9 Space-saving arrangement of joints in installation sites with restricted room

most stringent requirements in this respect are for lapped accessories for XLPE cables where high-insulation foils such as PE, cross-linkable PE and PP are used, as these materials show an increased tendency to develop a static charge and thus to attract particles.

The *spatial arrangement* of the sealing ends is largely invariable (insertion into existing transformer and switchgear housings or installation while adhering to the specified voltage-dependent flashover distances in an outdoor installation). As an example, Figure 7.8 shows the outdoor sealing ends of a 400 kV oil-filled cable. The spatial arrangements of joints must be appropriate to the circumstances on the site. In narrow joint pits, for example, it is advisable to offset the three joints as illustrated in Figure 7.9 in order to have sufficient space for installation.

The nature and extent of the installation operations are dependent on the cable construction, the accessories being used and the rated voltage. Details can be found in Section 6.

7.1.4 Completing the System

The final steps in the installation include installing additional components required for operation and – where there is an open trench – backfilling the trench and restoring the original surface. It has proved expedient to perform a check on the sheath or *corrosion protection* as described in Section 8.2.4 *before* filling in the cable trenches, to enable any sheath repairs that may be required to be carried out without further excavation work.

Depending on the cable type, system configuration and equipment, the following items can fall into the category of "additional components required for operation":

- Junction boxes for changing over the sheath connections at cross-bonding joints or for removing the ground connections at normal straight joints, to enable checks on the *sheath* or *corrosion protection* to be carried out (Section 8.2.4);

- Equipment for supplying low-viscosity impregnating media or compressed gas for paper-insulated cables, including the necessary pressure control and monitoring devices (Section 7.2);

- Cooling stations, heat exchangers and possibly separate cooling pipes for force-cooled systems (Section 5.1);

- Devices for measuring current and voltage, and (if appropriate) components for system monitoring – e.g. optical fiber cable (Section 7.4).

Further detailed information on laying and installation is provided in [148].

7.2 System Engineering

The system engineering varies considerably between paper insulated and polymer insulated cables, and therefore these two groups will be discussed separately below.

7.2.1 System Engineering for Paper-Insulated Cables

Paper dielectrics always require an impregnating medium at sufficient pressure to enable them to withstand high field strengths for long periods; depending on the type of cable, low-viscosity oil, high-viscosity or solid compounds may be used (see Sections 2.2.3 and 4.1). It follows that, depending on cable type, equipment for supplying impregnating medium or gas – including the associated pressure monitoring systems – falls into the category of essential components for this type of high-voltage and extra high-voltage cable. Differences arise from the type of impregnating medium used and the manner of generating pressure, as is demonstrated in the example of the system engineering for the three most widely-used types of paper-insulated cable (low-pressure oil-filled, internal- and external gas pressure cable).

Low-pressure oil-filled cable systems

The low-viscosity impregnating medium is connected via the hollow conductor to oil expansion tanks, inside which the necessary operating pressure is maintained by gas-filled diaphragm units as shown in Figure 3.7. The oil pressure is monitored continuously by *contact manometers* which trigger an alarm if the pressure falls below specified limit values; if appropriate these contact manometers will also de-energize the entire cable system to prevent breakdowns. Additional expansion tanks in conjunction with stop joints as illustrated in Figure 6.30 are needed when the system involves large height differences in order to limit the hydrostatic impregnating medium pressure. The system engineering of low-pressure oil-filled cable systems thus essentially comprises the components represented schematically in Figure 7.10.

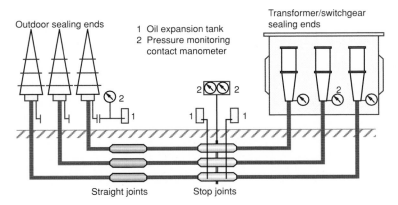

Figure 7.10 Essential components of a low-pressure oil-filled cable system

Internal gas pressure cable systems

Because the paper insulation of internal gas pressure cables contains extra-high-viscosity impregnating media that may be non-draining and which do not necessarily need to fill the available cellulose pores without leaving any unfilled spaces, there is no need to constantly change the liquid using reserve containers as with oil-filled cables. Instead, a constant supply of gaseous nitrogen at an operating pressure of 15 bar is required, along with the associated monitoring and alarm equipment.

However, when leaks occur this monitoring system generally does not automatically shut down the cable system when there is a risk of total *gas loss*. In fact the dielectric in internal gas pressure cables will withstand the operating voltage U_0 for a limited period of at least 24 h even when in a depressurized state. This is because gas loss at the core surface does not lead immediately to a pressure reduction within the cellulose pores that are not filled with impregnating medium; this effect is delayed in the context of a longer process of diffusion. Not until the positive pressure has dropped completely, in particular in the voids close to the conductor, is there any need to worry about lasting damage from corona discharges which would mean that the system needs to be de-energized. In the event of damage, the operator thus has sufficient time to first locate the cause of the problem, to dispatch engineers to the location of the damage and to carry out all the appropriate preparations for the repair before the system needs to be shut down. This considerably reduces the length of time that the service has to be interrupted, even in the case of more serious faults.

Other typical additional components of all three-phase gas pressure cable systems are the so-called *spreader heads* or *splitter joints*. In these accessories, the three conductors routed together in a steel pipe are separated at the cable end, initially into three separate conductors each with its own pressure-resistant, flexible metal sheath. Each of these conductors is then fitted with a separate sealing end.

In most cases the sealing ends themselves do not need any additional components with internal gas pressure cables. As with all high-voltage cables with a paper dielectric, the condensor cones (Section 6.2.2) used for stress controlling are generally housed in a pressure-resistant cast-resin insulator which is filled with an impregnating compound. Over a period of time a positive pressure builds up within the insulators, fed from the

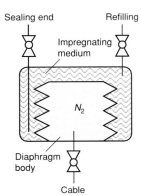

Figure 7.11
Principle of a pressure equalization tank for the gas-free generation of pressure for thermal stabilization of the sealing end insulation in gas pressure cables

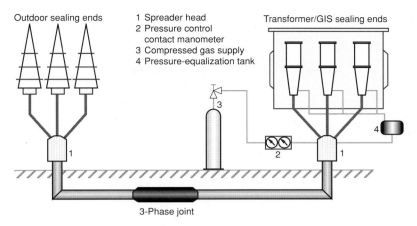

Figure 7.12 Essential components of an internal gas pressure cable system

cable insulation, and this is usually sufficient to thermally stabilize the sealing end dielectric (Section 3.1.2).

In special cases – e.g. in horizontally configured plug-in switchgear sealing ends – a shared *pressure equalization tank* (as shown in Figure 7.11) can be used for all three phases [150]. This takes advantage of the gas pressure which is already present in the cable system to apply a specified positive pressure *without gas* to the filling compound of the sealing end via a diaphragm unit, and at the same time to enable the non-compressible impregnating medium to undergo the necessary changes in volume due to the fluctuating thermal stress. Assuming the availability of a pressure equalization tank for supplying the switchgear sealing ends, a complete internal gas pressure cable system consists of the components that are represented schematically in Figure 7.12.

External gas pressure cable systems

In respect of the additional components required, internal gas pressure cable systems differ from their external pressure counterparts in just one aspect. Since, in this case, the compressed gas cannot penetrate to the dielectric and contribute to the thermal stabilization of the sealing end insulation, pressure equalization tanks as shown in Figure 7.11 are generally required for external gas pressure systems and for each individual sealing end (see Figure 7.13). Inside the (three-phase) joints shown in Figure 6.17 a similar solution is also used: In this case, too, a pressure equalization tank can be used to thermally stabilize the manually produced joint insulation, although it acts on all three phases simultaneously. As with all pressure equalization tanks, this is a totally maintenance-free component, and from the outside there is consequently no evidence of its existence in the joint housing.

External gas pressure cable systems are also configured in such a way as to allow them to continue to operate safely for at least 24 h at U_0 in the event of a pressure loss; there is consequently no need to shut down automatically when leaks occur. In this case, however, the reason for this capability is different from that with internal gas pressure cable, whose ability to remain operational in the event of gas loss was attributed to the fact that

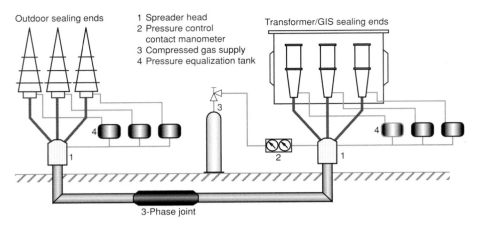

Figure 7.13 Essential components of an external gas pressure cable system

the pressure within the insulation drops only gradually: The dielectric, which at the moment the fault occurs is void-free and compressed, retains this electrically stable state at least until

- A pronounced expansion of the lead sheath initially occurs as a result of intense heating and

- The subsequent cooling-down process triggers a significant shrinkage in volume, especially in the impregnating medium, as a result of which voids capable of ignition develop in the dielectric.

Apart from the fact that this process requires at least one complete heating cycle as a result of the change in load, which takes virtually 24 hours simply because of the thermal time constant that applies to high-voltage cables, thorough studies point to the likelihood of even greater operational reserves. In fact dissipation factor measurements of the voltage in "pressureless" external gas pressure cables reveal that a number of heating and cooling periods pass before an ionization knee-point (as illustrated in Figure 3.6) develops as an indicator of the onset of thermal instability. This is evidently because of some kind of *memory effect* of the lead sheath. Because of these findings Siemens, for example, has promised safe continuing operation of its 110 kV gas pressure cable for at least 48 h after pressure loss at U_0.

7.2.2 System Engineering for Polymer-Insulated Cables

The absolute zero maintenance requirement of single-layer extruded dielectrics is frequently presented as a major advantage and so, in ideal circumstances, complete systems with polymer-insulated cables require no additional components (Figure 7.14). This, however, only applies as long as the accessories are also engineered accordingly. Most of the accessories for polymer-insulated cables that are described in Section 6.2.3 satisfy the zero-maintenance criterion. Only the back-to-back joint used in the extra high-voltage field, which is used in France, for example, and shown in Figure 6.26, needs a constant supply of SF_6 as an insulating gas, and can therefore no longer be classified as

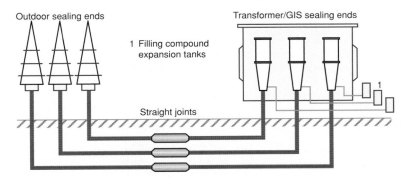

Figure 7.14 Components of a high voltage system with polymer-insulated cables

completely maintenance-free. The same point applies to the sealing ends for extra high-voltage cables that are likewise found mainly in France, and their slip-on stress cones are used to avoid surface discharges in insulating gas under increased pressure [133].

The sealing ends in widespread use for XLPE high-voltage and extra high-voltage cables in Germany, on the contrary, do not require any maintenance. It is here commonplace to embed the stress cones in suitable insulating liquids [132], and this forms a hermetically sealed system with no positive pressure and no external supply systems. Even if an *expansion vessel for the filling compound* is flange-connected, as in the case of horizontally configured switchgear or transformer sealing ends [151], the vessel does not need any additional maintenance. However, as an optional component of the cable system it is included in Figure 7.14 for the sake of completeness.

7.3 Electrical Interference from High-Voltage Cables

Conductors with current flowing through them and with a high-voltage potential are surrounded by an electromagnetic field whose influence causes voltages and currents to be induced in metal conductors of all types (data cables, instrument leads, telecommunication cables, but also gas and water pipes etc). In addition, fault currents can cause an increase in potential, e.g. at common ground connections, because of finite grounding impedances; for their part these ground connections can cause equalizing currents in other lines connected to the same ground. These effects, the intensity of which depends on a wide range of different parameters, are known as *electrical interference*.

Quite separately from this, the possible effects of power supply plant on humans and animals are assuming increasing significance within the intensifying debate about the environment. Electrical fields are of no significance in the context of high-voltage cables because they always have metal shields and/or outer sheaths; on the other hand, the magnetic field developed by the conductors that have current flowing through them is also still present above ground level in the vicinity of the cables, in areas accessible to members of the public. The height and pattern of this magnetic field with different system configurations are therefore considered briefly in a separate subsection (7.3.3) below, although the phenomenon in question is not electrical interference in the usual sense of the term.

7.3.1 Definitions

Before going on to discuss the different forms in which electrical interference appears, certain terms should be defined here which will aid understanding of the contexts in the subsections that follow.

- *Capacitive interference*
  This occurs only when an *electrostatic* field is present, and is therefore of no importance in the context of high-voltage cable systems because of the fact that they always have metal shielding.

- *Magnetic interference*
  This refers to the creation of an interference voltage in conductors caused by inductive coupling with high-voltage cables. This process has far greater practical significance than is the case with resistive and capacitive interference, and can occur both under normal operating conditions caused by the conductor current in the cable causing the effect, and – in a more intense form – in the event of a defect caused by a ground fault or short-circuit current. If the high-voltage cable lies parallel over a length l_p to a (conventional) telecommunication line (see Figure 7.15), the inductively coupled interference voltage U_i is calculated as follows [8, p. 351]:

$$U_i = I \cdot l_p \cdot \omega \cdot M' \cdot r \qquad (7.2)$$

  where
I	= Conductor current (fault current) in the high-voltage cable
l_p	= Length of parallel run between cable and line
M'	= Mutual inductance per unit length
ω	= Angular frequency
r	= Reduction factor (see below)

- *Resistive interference*
  This can be expected when high, usually rapidly-changing currents are flowing to ground across finite impedances and are thus causing an increase in potential in other lines or their screens with respect to the ground connections being used (Figure 7.16). Provided that the grounding is correctly configured, this process is only important in the event of a fault while, in comparison with magnetic interference, it remains virtu-

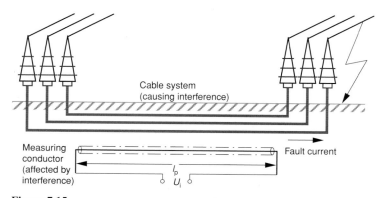

Figure 7.15
Determination of inductively coupled interference voltage in the event of a ground fault

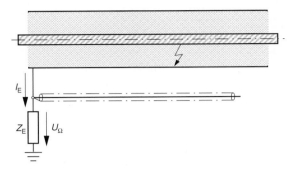

Figure 7.16 Determination of resistively coupled interference voltages

ally negligible. The level of resistively-coupled *interference voltage* U_Ω, e.g. in a telecommunication cable, is calculated in the same way as in Formula 7.2 [8, p. 358]:

$$U_\Omega = I_E \cdot Z_E \cdot r \qquad (7.3)$$

where

I_E = Proportion of fault current flowing across the grounding impedance Z_E
Z_E = Grounding impedance
r = Reduction factor (See below)

- *Long-term interference*
  Stationary or long-term interference is caused by intact cable systems with rated current and mains frequency. They have a long-term and almost uninterrupted effect on their environment.

- *Short-term interference*
  Transient or short-term interference, on the other hand, only occurs in the case of a fault, e.g. a ground fault or double line-to-ground fault. Accordingly it occurs only extremely seldom, and lasts no longer than a few seconds. Balanced against this is the fact that the intensities can be many times greater than with stationary interference.

- *Disturbance*
  Disturbance refers to a temporary adverse effect on the transmission of data, messages or measurement values, e.g. as the result of transient interference by power cables.

- *Reduction factor*
  The reduction factor is defined for faults (e.g. ground fault) and takes into account the fact that not the whole short-circuit current acts as interference. For example, in a ground fault, part of the current fed into the conductor from the transformer flows back to the transformer not through the soil, but through the cable sheath from the location of the damage (Figure 7.17). This sheath current compensates parts of the magnetic field developed by the affected phase, and thus reduces the magnetic interference produced. The generally applicable definition of the reduction factor r is as follows:

$$r = U_e / U_{e0} \qquad (7.4)$$

where

U_e = Amount of actual interference voltage
U_{e0} = Amount of interference voltage disregarding compensation

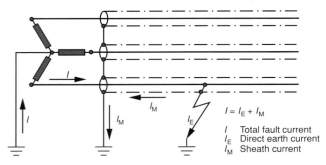

$I = I_E + I_M$

I Total fault current
I_E Direct earth current
I_M Sheath current

Figure 7.17   Defining the reduction factor

The resulting reduction factor is a real number with a value between 0 (no interference whatsoever) and 1 (no reduction whatsoever). For the example of the ground fault (Figure 7.17) with the partial currents (values) entered there, r is calculated as follows:

$$r = I_E / I = (I - I_M) / I \tag{7.5}$$

If a circuit configuration includes several compensating elements (e.g. a conductive screen on the affected line as well as the metal sheath of the interfering high-voltage cable), then the resulting reduction factor r_{res} is produced from the product of the separate factors:

$$r_{res} = r_1 \cdot r_2 \cdot r_3 \cdot \dots \tag{7.6}$$

- *Compensating conductors*
  Metal lines (overhead ground wires, rails, pipes etc) in the vicinity of a high-voltage cable, through which the interfering effect of the latter is reduced in accordance with Formula 7.6.

- *Expectation factor*
  This is always ≤ 1, and in certain cases is used as a factor for multiplying an interfering short-circuit current. This approach takes into account the statistic that the likelihood of all adverse circumstances being combined in a single fault situation is very small. So, for example, it is highly unlikely that the short-circuit current will flow along the entire interference length [8, p. 352].

7.3.2 Possibilities for Influencing Electrical Interference

The possibilities for affecting electrical interference caused by high-voltage cables will only be considered below in terms of magnetic interference, which is the most important case both from the technical and the practical point of view. Further information on the subject of interference can be found in [8] and [42].

As an example, let us consider an arrangement which is encountered frequently in urban power supplies (Figure 7.18, [152]): a 110 kV cable system with a flat arrangement runs parallel to a pipeline supplying gas, water or heating. In accordance with Formula 7.2

$$U_i = I \cdot l_p \cdot \omega \cdot M' \cdot r$$

the intensity of magnetic interference of a cable on the parallel pipeline at a constant frequency (50 Hz; $\omega = 100\pi$ s^{-1}) is determined by the level of inducing current I, the geo-

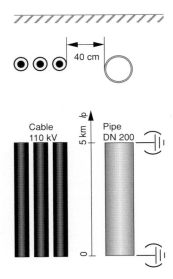

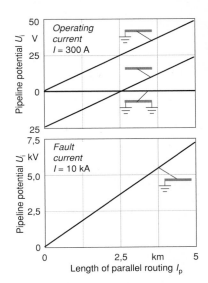

Figure 7.18
Layout for estimating the magnetic interference caused by high voltage cables on pipelines (as described in [152])

Figure 7.19
Interference voltages on the pipeline as illustrated in Figure 7.18 depending on the current

metric proportions of the arrangement (length of parallel routing l_p, mutual inductance M) and the reduction factor r. All the parameters here are directly proportional to the level of the inductive interference voltage.

The absolute voltage level occurring on the pipeline with respect to ground potential is additionally determined by the grounding conditions of the pipe, as shown by Figure 7.19. Here, the worst-case situation is with grounding at one end only; in this case, an interference voltage develops at the free end of the pipe that is exactly twice the maximum value in the potential-free state. Grounding both ends prevents, by definition, the occurrence of an interference voltage relative to ground but, on the other hand – like both-sided grounding on non-transposed cable sheaths (Section 3.4.1) – causes equalizing currents along the pipe combined with additional resistance losses and possibly occurrence of hysteresis.

An important fact emerging from the voltage patterns shown in Figure 7.19 is that there is no reason to be worried about dangerous interference to potential-free pipelines in the vicinity of high-voltage cables in normal operation and under realistic conditions. Only in the event of a fault with ground fault currents of several kA can voltages of several kV be induced briefly on pipelines that run parallel and close to the cable system for a considerable distance.

Given these conditions, the reduction factor that can be formed in accordance with Formula 7.6 by combining several multipliers becomes more important. The single most important factor is the current reduction factor r_{Ki} of the interfering cable itself, as it is always present. It is calculated using the fault example in Figure 7.17 via Formula 7.5

$$r_{Ki} = I_E / (I_E + I_M)$$

to $\quad r_{Ki} = \sqrt{\dfrac{R_M^2 + (\omega L_M)^2}{(R_M + R_E)^2 + \omega^2 (L_M + L_E)^2}}$ (7.7)

where R_M, R_E are the ohmic sheath resistance and ground resistance respectively

$\quad L_M$, L_E = Inductance of sheath or ground circuit

$\quad \omega \quad$ = Angular frequency

In practice the ohmic ground resistance and, in the case of non-magnetic sheath materials, the inductive sheath resistance relative to the ohmic resistance, can generally be disregarded [8, p. 352]. The sheath inductance is also considerably less than that of the ground circuit, thus:

$$R_E, \ \omega L_M \ll R_M \text{ and } L_M \ll L_E$$ (7.8)

so that the expression for the reduction factor can be simplified to:

$$r_{Ki} = \dfrac{R_M}{\sqrt{R_M^2 + \omega^2 L_E^2}}$$ (7.9)

Since the impedance of the ground circuit in a fixed system configuration cannot be altered, the only opportunity to reduce the resistance factor for the cable itself lies in the use of low-resistance cable sheaths and/or shields. This includes using more conductive sheath materials as well as increasing the sheath cross section, as be seen from Figure 7.20. It shows how the reduction factors change according to the cable diameter for various 110 kV single-conductor cables with lead and aluminum sheaths and copperwire screen laid in flat formation, and for a three-phase gas pressure cable with a steel pipe.

In the directly comparable metal sheaths of oil-filled cables, lead produces a two to three times greater reduction factor because of its lower conductivity; in both cases the factor is reduced as the sheath diameter increases because of the associated increase in cross section. On the other hand, a slight increase in reduction factor over and above that associated with the cable diameter can be seen in XLPE cable with a copper wire screen. The

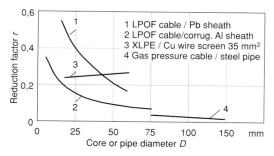

Figure 7.20　Reduction factors of various 110 kV cables (as described in [152])

reason for this is that the cross section of copper in the wire screen is kept constant (35 mm^2) irrespective of the diameter, along with the slightly changing inductive impedance ratios. In comparison with the single-conductor cables, the three-phase steel pipe cable has a still better (smaller) reduction factor, based on the system configuration, which is advantageous in this respect (Table 7.2) combined with the necessarily large cross section of the pressure-resistant pipe.

The situation turns out to be considerably more complicated when there is ferro-magnetic armoring with a relatively small cross section, in which magnetic saturation effects can develop under the influence of large fault currents. Granted the assumptions above, the reduction factor then becomes current-dependent, since the sheath inductance L_M in Formula 7.7 falls as the degree of saturation increases. As an example of such behavior, Figure 7.21 shows the reduction factor patterns in a steel-tape armored lead-sheathed cable of different dimensions depending on the inductance current.

Apart from using low-resistance sheaths, there are a number of other possibilities for reducing interference by high-voltage cable systems or the consequences of interference on lines in the vicinity. Along with system configuration changes as shown in Table 7.2 (e.g. trefoil rather than flat formation, reduction of the distance between phases), these also include e.g.:

- The use of additional *compensating conductors* in the form of conductors grounded on both sides between the cable system and the line subject to interference and

- The use in low-current lines of coaxially shielded conductors rather than just twisted conductors.

This leads in both cases to a reduction in the resulting reduction factor as defined in Formula 7.6 in the form of additional multiplicands <1. With all the options discussed here for influencing electrical interference with the aim of reducing it, it is important to bear in mind that positive changes generally have a negative effect on other operational characteristics of the cable system, and especially on sheath losses (see Table 7.2). The aim for project planners must therefore be to achieve the optimum characteristics while at the same time bearing the investment costs in mind.

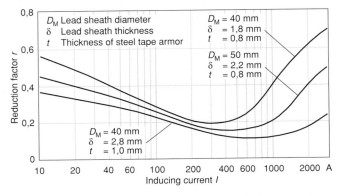

Figure 7.21 Reduction factor of single-phase lead-sheathed cables with steel tape armor [8]

7.3.3 Magnetic Field at Ground Level

The possibility that mains-frequency electromagnetic fields produced by high-voltage systems could have adverse effects on health were first widely discussed within the profession during the 1970s, instigated by a Russian Cigré article [153]. This formed the starting point for a number of specific basic studies in this field (e.g. [154 – 156]). Largely taking as their starting point the conditions below high-voltage overhead lines, these studies essentially concentrated on the electrical field [157]; the magnetic field was largely ignored [158].

Meanwhile, however, weak magnetic fields – such as those which can occur at ground level directly above a cable system – are increasingly entering the debate [159]. In order to establish a sound starting point, Cigré commissioned the 'joint task force 36-01/21' to work out suitable calculation methods for determining the magnetic fields caused by variously configured cable systems, and comparing their results with on-site measured values. An initial report on this research was published in 1996 [160], and a further report is in preparation. In addition, maximum magnetic induction or flux density values were incorporated in national and international standards, ranked according to the duration of exposure. They were then restricted still further by the *German Federal Law for Protection against Emissions* that was in force at the time (see Table 7.4)

For these reasons a brief consideration of certain aspects of the magnetic compatibility of high-voltage and extra high-voltage cables follows below. The main area of interest here is long-term operation, unlike in the context of interference with parallel-routed lines. Faults, on the other hand, do not give rise to fears of damaging consequences since – despite the far greater field strengths involved – the time for which these are effective is extremely short.

A series of wide-ranging mathematical and experimental studies applied to high-voltage and extra high-voltage cables were carried out by a cigré task force [160]. When the system is configured as a single system (flat formation with a distance of 30 cm separating the phases and laid at a depth of only 0.5 m (an unfavorable system configuration in this respect), a *magnetic induction (flux density)* of only about 45 µT/kA is set up directly above the cable system, and this also fades very rapidly either side of the cable system. Figure 7.22 gives an idea of the conditions. All other practical system configurations, in particular the trefoil formation of phases or laying at a greater depth as described in Section 7.1.1., lead to a further marked reduction in the magnetic field at ground level emanating from the cable system (see Figures 7.22 and 7.23).

Table 7.4
Max. permissible magnetic induction values in standards and regulations

Duration of action	1h / d	2h / d	8h / d	Constant
DIN VDE 0848 Part 4 A3	4,240 µT	2,550 µT	1,360 µT	424 µT
ENV 50166–1	No data mentioned	No data mentioned	1,600 µT	640 µT
German Federal Law for Protection against Emmissions	No data mentioned	No data mentioned	No data mentioned	100 µT

Although in the worst case this field is added directly to the earth's magnetic field, only inductance with an order of magnitude of < 100 µT (the tolerance value specified in the German Federal Law for Protection against Emissions) can arise directly from high power cables. In order to bring about a further reduction in the magnetic field at ground level without restricting the transmission capacity, a whole range of different measures can be adopted – as in connection with electrical interference (Table 7.2). These can be divided into three categories:

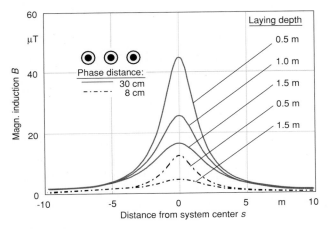

Figure 7.22
Profiles of magnetic induction above high voltage cable systems with flat formation (as described in [160])

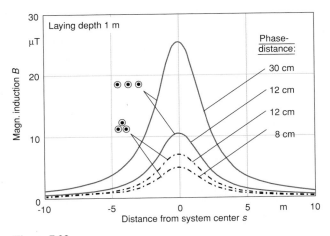

Figure 7.23
Profiles of magnetic induction above high voltage cable systems with various different configurations (as described in [160])

233

- *Improved laying conditions.* This category includes
  - Optimized system configuration (trefoil rather than flat formation),
  - Reduced gap between phases and
  - Increased laying depth
- *Ferromagnetic shielding* by
  - Covering cables with steel plates or
  - Pulling into steel pipes.
- *Compensation of the magnetic field* by
  - Optimized phase sequence in double and multiple systems (Figure 7.24) or
  - Installing compensating conductors above the cables (Figures 7.25 and 7.26).

However, here too the majority of the measures described – like the options for restricting interference – are also associated with a more negative aspect, in the form of the loss balance or heat dissipation of the cables (see Table 7.2). So in this case, too, all the advantages and disadvantages must be considered carefully.

7.4 Monitoring

In the context of cable systems and other electrical power supply systems, *"monitoring"* describes methods of measurement that enable particular operating statuses and properties to be watched either regularly or continuously without interrupting the power supply (*"online monitoring"*). The objectives of cable monitoring include preventing damage, increasing availability and/or reducing the down time of systems, and optimizing their

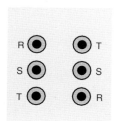

Figure 7.24
Optimized phase sequence in a double system for reducing the overground magnetic field

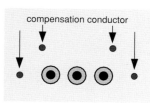

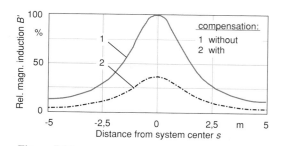

Figure 7.25
Compensating conductors for reducing the overground magnetic field

Figure 7.26
The effects of compensating conductors (as illustrated in Figure 7.25) on the magnetic field

234

loading. Table 7.5 provides a brief summary of some of the measured variables discussed for online monitoring, their purpose and anticipated usefulness for users. Subsections 7.4.1 – 7.4.3 below contain additional details on the technical, functional and economic aspects of three methods which have been chosen as examples, and which are already widely available for high-voltage and extra high-voltage cables:

- Cable temperature monitoring

- Detection of water in the cable

- Measurement of partial discharges in polymer-insulated cables and/or their accessories.

On the other hand, no further details are given of two other important measured variables which have been continuously monitored online without interrupting the power supply ever since cable engineering began: operating current and operating voltage. Current and voltage measurements obtained with the help of suitable transformers should probably also be included in the 'monitoring' category in the sense of the definition above.

Table 7.5 Methods of monitoring for cable systems

Measured variable	Objectives	Development status	Bibliography
Axial distribution of cable-temperature	• Detection of hot spots • Prediction of temperature rise pattern on overload • Optimum thermal usage	In use	[161–164]
Water penetration under the sheath of polymer-cables	• Location of leaks • Prevention of water treeing and screen corrosion • Replacement for sheath testing	Laboratory development completed, product development being prepared	[164–167]
PD monitoring in accessories for polymer cables	• Early detection, location & evaluation of accessory faults • Planning of shutdowns and repairs • Damage limitation	Ready for production, practical use being prepared	[164,168,169]
PD monitoring of cable systems having polymer-insulation	• Early detection, location & evaluation of faults • Planning of shutdowns and repairs • Damage limitation • Replacement for site testing	Research & development	[164,170,171]
Leak monitoring of oil-filled-cable systems	• Prevention of soil contamination • Planning of shutdowns and repairs • Easier approval	Laboratory development completed	[164]
SF_6 monitoring in oil-filled cable systems	• Early detection & evaluation of leaks in switchgear sealing ends • Planning of shutdowns and repairs • Damage limitation	Ready for production	[172]

7.4.1 Cable Temperature

The load capacity of underground cables can be severely compromised by factors such as the conditions in the surrounding soil, by so-called congestion points, i.e. the proximity of other cable systems, district heating lines, and by the adversely affected dissipation of heat in the area surrounding an accessory (Section 3.5). The problem here is that the influencing factors are local and unevenly distributed and their effects cannot be calculated with precision. The only option until now for avoiding premature ageing due to local overheating has been to remain slightly below the maximum capacity calculated theoretically for the sake of safety [152]. In this way, though, the full potential of the cable is generally not exploited, or it is over-dimensioned to a certain extent from the outset.

The situation is improved by using on-line monitoring of the actual axial temperature distribution in the area around the screen or even in the conductors of the separate cables using *optical fibers* in conjunction with suitable analysis software [161, 162]. The measuring method is based on the feature that is especially pronounced in *multimode gradient optical fibers*, namely that light impulses of a specific wavelength that are sent out at the start of the optical fiber are reflected with varying intensity according to the temperature (*Raman effect*) [163]. *Instruments for measuring backscatter* which are set up for this function supply a temperature profile along the whole length of the cable, and their results not only establish locally prevailing temperatures to $\pm 1\,°C$, but also enable the individual values to be located with a precision of within 1 m as long as the cable length is no more than 10 km.

Optimum precision is achieved by using two optical fibers of the same kind, which are welded together (i.e. optically short-circuited) at the far end of the cable. The backscatter measuring instrument thus supplies two mirror-image temperature profiles from which the position and height of individual peaks can be pin-pointed with greater precision through interpolation. Figure 7.27 shows the theoretical layout of this type of double optical fiber system, and Figure 7.28 shows the mirror-image temperature profile of a 200-m-long XLPE high-voltage cable that has been plotted using such a system [164].

The fiber optic sensors required for monitoring are generally enclosed in a small tube made from high-quality steel, and are thus mechanically and thermally stable enough to withstand the stresses encountered during installation in the cable conductor or screen and during the cable laying operation. For cables with no wire screen, a new type of sensor-conductor with a non-metallic sheath has been developed, and this can, for example, be easily integrated into the padding material of an XLPE cable with a lead sheath. Actual examples of optical fiber under the sheath of XLPE high-voltage cables with/without a wire screen can be seen in Figure 7.29. In areas close to joints the sensors of the two jointed cables are linked together and there are suitable techniques available for bringing the optical fiber out for the various different types of sealing ends. Figure 7.30 shows as an example the terminal boxes for measuring temperature on plug-in switchgear sealing ends of XLPE high-voltage cables.

In the simplest case, the online monitoring of cable temperature provides exact information about the temperature condition prevailing at any given time and any given location on a transmission run; this information enables the operator to utilize the system right up to its thermal limits. In combination with intelligent software ("*load monitoring*" [162]), the opportunity also exists to use the change in monitored temperature over a period of

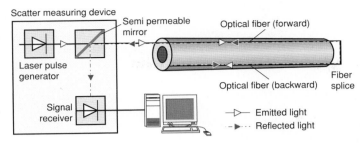

Figure 7.27
The principle of a monitoring system for measuring temperature in high voltage cables using two optically short-circuited optical fibers [164]

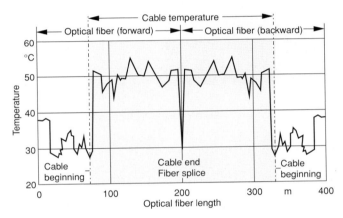

Figure 7.28
Mirror-inverted temperature profile along a 200-m-long cable, using a monitoring system as illustrated in Figure 7.27 [164]

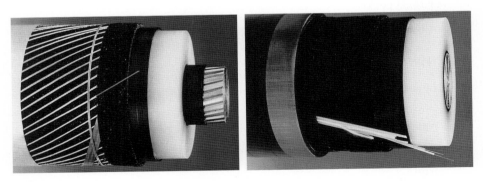

Figure 7.29
XLPE high voltage cable with temperature sensors. Left: Optical fiber integrated in wire screen, right: Optical fiber under lead sheath

237

Figure 7.30
Switchgear sealing ends with optical fiber terminal boxes for measuring temperature in XLPE
high voltage cables

time to predict the permissible duration of the overload state in the case of temporary overloads. It is also possible, if the time of a load peak is known, e.g. as a result of maintenance work on parallel systems, to determine in advance the maximum permissible transmission capacity during this high-load period. Unlike conventional programs concerned with temperature in cables under transient load conditions, 'intelligent' monitoring also takes into account the current environmental conditions along the section of cable concerned when making a prognosis. Thus, for example, it will take into account the actual additional thermal load currently emanating from adjacent sections at particular congestion points.

Temperature monitoring thus contributes to better usage of the system, helps with the advance scheduling of maintenance work, and even sometimes helps with the avoidance of outages when faults occur. The extent to which this justifies the increased cost involved in the manufacture and installation of cable systems with integral optical fiber sensors must be determined for each individual case at the system planning stage.

7.4.2 Water in the Cable

If not earlier, then at least since the increasing number of failures of PE/XLPE-insulated medium voltage cables due to water treeing [88], it has been recognized that water must be largely kept away also from the polymer insulation. Modern XLPE high-voltage cables are therefore generally provided with an enclosed metal sheath (Table 2.4). Occasional leaks in this moisture barrier caused by mechanical effect or corrosion cannot, however, be avoided. In order to prevent resultant damage to the insulation, such leaks should be detected and rectified as soon as possible.

The only way of detecting sheath faults in existing cable systems without any special equipment is to carry out a *corrosion-protection test* under moderate DC voltage. However, there is a considerable cost associated with this approach, which also, as a rule, requires that the transmission section concerned is disconnected from the network (Section 8.2.4).

Sensors have been developed recently specifically for polymer-insulated high-voltage cables which not only enable the existence and location of water penetration under the sheath to be declared with greater certainty than is possible with a corrosion protection test, but also – in their role as continuous online monitoring devices – report faults almost at the moment they arise [164]. Two different measuring principles can be distinguished here according to the type of *water sensors* integrated into the shield zone as shown in Figure 7.29 [165]:

- Optical measurements using optical fiber (e.g. [166]). As in temperature monitoring, they take advantage of the characteristic of the glass fibers used which makes them either scatter or reflect light more strongly by themselves or in conjunction with a suitable sheath when affected by moisture. However, the techniques involved are still largely at the research stage.

- Electrical measurement using metal sensors with moisture-sensitive insulation [164, 167]. In this case the crucial variables measured are the electrical resistance and a voltage or current between sensor-conductor and cable screen, or between two sensor-conductors. Systems of this type have now reached the stage of practical trials; one example is explained below in slightly greater depth.

In the simplest case the measurement circuit for monitoring moisture consists of a metal wire with a moisture-sensitive sheath integrated into the cable screen, a voltage source and a measuring resistance with parallel-connected voltmeter for measuring current. The voltage source and measuring resistance are connected at the end of the cable between the sensor wire and the screen or metal sheath as shown in Figure 7.31.

In the dry state the sensor and screen/sheath are insulated from one another; there is no current flow. On the other hand, if any moisture has penetrated into the screen area at any point, then the sensor sheath loses its insulating effect at this point apart from a finite contact resistance R_K. The value of current that can be measured at the cable end as an indicator of water damage is influenced by a number of factors apart from the value of the driving voltage, including the (undetermined) size of the contact resistance and the

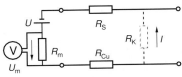

R_K Contact resistance
R_S Sensor resistance
R_{Cu} Screen/sheath resistance
R_m Measuring resistance

Figure 7.31
Mode of operation of an electrical water sensor

239

line resistances between the cable end and the point of damage. The simple sensor circuit as in Figure 7.31 is therefore capable of detecting damage, but not of locating it.

Location of a fault, on the other hand, generally requires a measurement from both ends and the use of sensor-conductors of a suitable size with a specified resistance layer. Providing that is the case, then two current circuits result in the event of damage as identified earlier by the occurrence of a finite contact resistance R_K. The line resistances of the circuits, R_{Si} for the sensor wire and R_{Cui} e.g. for the copper screen of a laminated-sheath cable are directly proportional to the distance of the point of damage from the respective ends of the cable (i = 1 or 2):

$$R_{Si} = R_S' \cdot l_i \tag{7.10}$$

and

$$R_{Cui} = R_{Cu}' \cdot l_i \tag{7.11}$$

where

R_S', R_{Cu}' = Resistance per unit length of the sensor wire or copper shield in Ω/m
l_i = Distances between the cable ends and the point of damage in m

Due to the existence of two current circuits with the common circuit element R_K, the undefined contact resistance of the point of damage, it is possible to form two voltage-loop formulae, from which the contact resistance can be eliminated. On the (reasonable) assumption that the supply voltages U and measuring resistances R_m are identical at both ends of the cable, the so-called *location formula* for the electrical water sensor as shown in Figure 7.32 is eventually produced in association with Formulae 7.10 and 7.11 [165]:

$$l_1 = l_K \cdot \frac{I_2}{I_1 + I_2} - \frac{R_m \cdot (I_1 - I_2)}{(R_S' + R_{Cu}') \cdot (I_1 + I_2)} \tag{7.12}$$

where

l_1 = Distance between cable end 1 and the point of damage in m
l_K = Total length of cable in m
I_1, I_2 = Currents in circuits 1 and 2

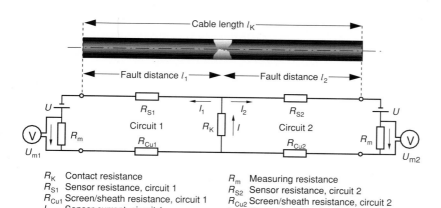

R_K Contact resistance
R_{S1} Sensor resistance, circuit 1
R_{Cu1} Screen/sheath resistance, circuit 1
I_1 Sensor current, circuit 1
U_{m1} Measuring voltage, circuit 1

R_m Measuring resistance
R_{S2} Sensor resistance, circuit 2
R_{Cu2} Screen/sheath resistance, circuit 2
I_2 Sensor current, circuit 2
U_{m2} Measuring voltage, circuit 2

Figure 7.32 Equivalent circuit diagram of an electrical sensor for locating water damages

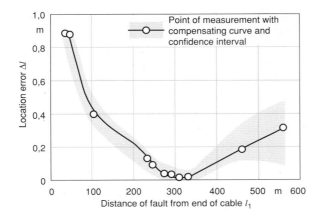

Figure 7.33
Location error Δl when locating sheath failures involving ingress of water in a 600-m-long XLPE cable (as described in [164])

The required distance of the point of damage from the cable end is thus linked back to data or measured values, all of which are known. Nonetheless, the determination of the location is adversely affected by disturbance variables. These include most notably the voltage induction resulting from the conductor currents in measurements with mains-frequency alternating voltage and the formation under the effect of moisture of a *galvanic element* between sensor-conductor and cable shield/metal sheath, as a result of which measurements with direct voltage can be distorted. In the interest of precise location a number of parameters must therefore be optimized:

- Resistance per unit length of sensor-conductor not too high ($R_S' < 10 \ \Omega/m$);
- Applied voltage not too low ($U \geq 5$ V);
- DC voltage used with positive polarity at the cable shield;
- Measuring resistances R_m and actual level of voltage applied U appropriate to the cable length l_K and resistance per unit length of sensor R_S'.

In this way the *location precision* with electrical moisture sensors can be increased to around 99.9%, or the *absolute margin of location error* Δl irrespective of cable length and location of fault is no more than 10 m. Even the most minute amounts (<1 ml) of water at the sensor will produce a measurable signal [164].

As an example of the precision that can be achieved, Figure 7.33 shows the *location error* Δl relative to the distance l_1 between the end of the cable and the location of the fault for measurements on a 600-m-long XLPE cable with a laminated sheath and copper screen. The absolute error defined thus will always remain below 1 m; relative to the total length l_K of the cable this produces a *relative location error* $\Delta l / l_K$ of no more than 1.5 ‰, with the largest discrepancies arising towards the ends of the cable.

7.4.3 Partial Discharge Monitoring

The threat to PE/XLPE-insulated high-voltage cables and their accessories posed by partial discharges (which was established in Section 3.2.1 amongst others) is responsible for the fact that every section of cable supplied and also an increasing percentage of prefabricated accessories are subjected in the context of component tests to a sensitive partial discharge measurement with an increased alternating voltage (in the case of the cables, the PD test has now been mandatory for many years since the introduction of standards to that effect – see Section 8). For this purpose most producers have set up fully electro-magnetically-shielded test laboratories that make it possible to perform PD measurement with great sensitivity with an order of magnitude of 1 or 2 pC.

Meaningful on-site partial discharge tests have so far mostly failed because of the high *noise levels* from high-frequency radio signals, switching pulses and corona interference from neighboring sections of the network. These overlay any PD pulses that may be present in the cable being evaluated – often with several times the amplitude.

Not until fairly recently have promising approaches been developed which, in the future, will enable cable systems with appropriate supplementary equipment to undergo sufficiently sensitive PD tests on site, at least in sections [164, 168-171]. The degrees of progress made in the various different developments are still very varied, according to the respective objectives. Developments can be divided into three categories:

- PD *measurement* and PD *location,*
- PD *testing* and PD *monitoring,*
- Inclusion of the whole *cable system* or only the *accessories.*

According to the current state of the art, true PD monitoring in the sense of the definition provided at the beginning of Section 7.4 (without any interruption to power transmission) will not be available in the foreseeable future for large cable systems. Instead it will initially only be available for the accessories, particularly joints, and only where they are equipped with suitable sensors. So-called *directional couplers* are useful in the sense of largely eliminating disturbances that are fed into the cable system from outside. They must be installed as close as possible to the accessory to be monitored (in the case of joints on both sides), and they detect the direction from which a pulse is arriving. In this way interference pulses from the cable section can be distinguished from partial discharges from the joints, and excluded from further evaluation in the measurement system. With the addition of digital filtering of continuous sine-wave interference signals (e.g. from radio transmitters) and the use of pattern-recognition algorithms to eliminate statistically pulse-shaped sources of interference, partial discharges can at least be detected in an unshielded environment with an order of magnitude of 1 pC and possibly even attributed to individual joints.

From a technical point of view there are two solutions of more or less equal merit in the frame for the direction-dependent detection of partial discharge pulses:

Figure 7.34 Rogowski coil

- *Inductive couplers*

  Inductive couplers (so-called *Rogowski coils* [171]) consist of a ferro-magnetic annular core with winding (Figure 7.34), which concentrically encloses the cable to be monitored. For the directional effect to work effectively, two identical Rogowski coils must be used, and they must be wound in *opposite* directions. Their output pulses are first combined to form a summation signal which is then analyzed further. Figure 7.35 clarifies the mode of operation. Pulses from the joint that is being monitored pass through both Rogowski coils in *opposite* directions and, in summation, produce a signal with almost twice the amplitude because the winding on the coils is likewise in opposite directions. For the same reasons pulses that originate from the right or left of the joint and are consequently passing through the Rogowski coils in the *same* direction are largely canceled out during summation. The advantage of this inductive PD decoupling lies in the opportunity of constructing the annular cores of the Rogowski coils from two fold-out halves, thus enabling joints to be easily upgraded even after their installation underground.

- *Capacitive couplers*

  Acording to figure 7.36 capacitive couplers consist of ring electrodes arranged concentrically around the undamaged outer semiconductive layer, each with two measuring taps linked to the electronic analysis equipment. At sufficiently high frequencies, the conductive layer loses effect (becomes non-conductive), and the ring electrode forms a short cylindrical capacitor with the cable conductor. In addition, as the frequency increases, so the inductive impedance of the electrode arrangement also makes its presence felt increasingly. It is thus possible to implement directional detection for PD pulses with an appropriate combination of electrode geometry and measurement frequency through the exploitation of runtime effects [168]. The selectivity of this solution is comparable to that of the two Rogowski coils, although the installation cost associated with the joints is considerably less. For this reason Siemens has equipped the joints in the 400 kV XLPE cable system "Bewag III" in Berlin [105, 127] (which has been finished, succesfully tested and energized in 1998) with a PD monitoring system using the capacitive directional coupler sensors described above.

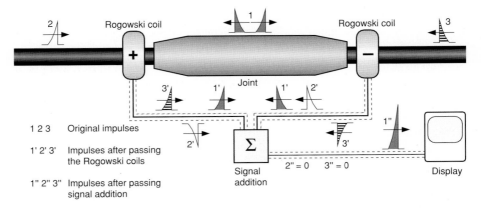

Figure 7.35
The mode of operation of inductive directional couplers (Rogowski coils) for detecting partial discharges from joints [171]

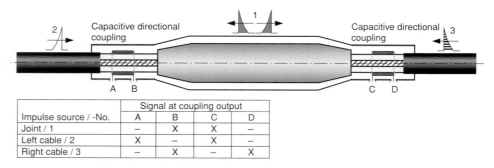

Impulse source / -No.	Signal at coupling output			
	A	B	C	D
Joint / 1	–	X	X	–
Left cable / 2	X	–	X	–
Right cable / 3	–	X	–	X

Figure 7.36
Arrangement and output signals of capacitive directional couplers for detecting PD from joints [168]

Irrespective of the type of tap, the decoupled pulses in the immediate vicinity of the monitored accessory are converted to optical signals, fed into an *optical fiber bus* running parallel to the cable system and thus passed, free from electromagnetic interference, to an analysis station in the control room [164].

In contrast to the sensitivity of PD monitoring on accessories at around 1 pC, with online monitoring of complete cable systems in network operation, the best result achieved so far under optimum conditions has been a resolution of around 10 pC and, under favorable conditions, even values significantly in excess of 100 pC are recorded [171]. Better results can be achieved using alternative forms of loading, such as damped oscillating or unipolar voltage pulses with decay times of several ms [169]. Their application, however, means the abandonment of the (interruption-free) monitoring principle; a more detailed appraisal of this technique can therefore be found in Section 8.3 (PD tests on polymer-insulated cables).

In summary, this means the following: meaningful PD monitoring on extensive polymer-insulated high-voltage cable systems has been limited to date to joints that have been equipped specifically for that purpose and to sealing ends that have been fitted with suitable sensors. Both capacitive and inductive couplers may be suitable and, if possible, they should display directional transmission characteristics as illustrated in Figure 7.35.

There is still a long way to go from this actual situation to a comprehensive *system diagnosis* while continuing operation, and intensive research and development will be required. It is not just a case of perfecting the monitoring techniques but also and, most importantly, of correctly interpreting the data that will then be available "online". In particular, the question of the degree of ageing and the remaining service life of polymer insulation that can be deduced from it may only be possible to answer with considerable reservation – even when all the technical measurement data that can possibly be acquired are available. This is illustrated by the difficulties involved in analyzing test results in the so-called *life diagram* shown in Figs 3.13 and 3.14 in Section 3.2.2 (for further details see Section 8.2.1, "Development tests").

8 Tests on Cables and Cable Systems

8.1 Quality Management

The competitiveness of a company is to a large extent determined by the quality of its products, services (e.g. installation) and processes. An essential prerequisite for this is an effective and comprehensive quality management (QM) system.

Customers today expect the company manufacturing their products to support this type of QM system, and also to maintain up-to-date certification. This gives them the security of knowing that all personnel bearing responsibility for the relevant processes are known, and their authority is controlled. The customer can then safely assume that the necessary pre-requisites for fulfilling their order, and thus also for the overall *quality of the product* are in place.

8.1.1 Quality Control

The term *testing* is defined in national and international standards and specifications as a visual or measured comparison of specified parameters or requirements with the actual values of a product. The quality of the product is a measure of the extent to which the various requirements and therefore also customer expectations have been fulfilled. Quality is thus ultimately defined by the demands of the customer and/or the marketplace.

The term quality was for a long time linked purely with a reflection of technical aspects; quality was defined only in terms of the product. But requirements nowadays encompass not only the technical aspects but also factors relating to the environment and usage as well as the legal and economic sides.

When the only quality tests or checks are carried out at the end of the production process, it is only possible to demonstrate whether or not the technical requirements have been fulfilled, or where there are deviations from the specifications. Studies have shown that around 75 % of the deviations revealed in final checks were brought about much earlier in the development or design stages of the product development process, i.e. before commencing production. When errors arise at this early stage and are then "refined" during the different production stages that follow, this has a considerable effect on timescales and costs; the result is reworking, repair and re-manufacture.

8.1.2 Quality Assurance

On the production line, the products are created in a *value creation chain*, which is made up from a number of different processes. Typical processes forming this chain are order processing, development, product design, procurement of materials, production. If one starts from the view that error-free products can only be created by error-free processes, then all the processes in the chain must be developed in such a way as to avoid any unintentional errors. In process-oriented sequences this is achieved by integrating testing activities as sub-stages in the processes (self-responsible quality assurance at the workplace). This extends the concept of quality from the product to the "process" and the service provided by the item in question.

In contrast with the quality control and final control which preceded it, so-called *quality assurance* takes the process into a new dimension; the identification of errors is replaced by the avoidance of errors. Quality assurance extends through the entire process of product development, from development through manufacture to installation and commissioning. The quality of the product is no longer simply tested, instead it is worked out from the quality of work in the individual processes in the development chain that are carried out by man and machine. The following two principles are of paramount importance:

- The quality of the product arises from the quality of work, and

- Individual quality forms the basis for overall quality.

In accordance with the above principles, quality is achieved through precautionary measures that give absolute control over all the processes that are required in the organization, and is the responsibility of all personnel. In this system there is no position in the company that does not have tasks and objectives which influence quality. The quality of all relevant business processes thus becomes a pre-requisite for the quality and reliability of the products developed, produced and sold by a manufacturer.

For companies operating in the international market the standards produced by the International Organization for Standardization (ISO) are of great importance in this respect. ISO has created a series of international standards which is presented in Germany under DIN EN ISO 9000 et seq. The norms in the series ISO 9000 et seq. have set standards for quality management systems, and have been introduced in many different economic sectors around the world.

DIN EN ISO 9000 describes the systematic structure of a self-contained QM system. A system of this type also builds a continuous improvement system into the company, and above all aims to ensure product quality by detecting in advance and avoiding sources of errors. This regulatory work sets up requirements for every step in the development process of a product, starting with the sales activities and extending right through to delivery and commissioning (e.g. of a cable system). The points below are extracted from the standard with the intention of providing examples to outline the procedure.

- *Design control*
  The development and design of a product is achieved through comprehensive and clear definitions of objectives, appropriate documentation of the stages of development and the checking of results by qualified personnel.

- *Process control*
  All stages of production are planned in advance and specified in writing in order to ensure that they take place under controlled conditions.

- *Control of measuring and testing equipment*
  All measuring and test equipment must be periodically calibrated, adjusted and maintained.

- *Corrective* and *preventive measures*
  These measures are designed to rectify the causes of errors and to eliminate repetition of errors.

- *Internal quality audits*
  The effectiveness of the QM system must be regularly checked and confirmed.

A QM system based on this code offers the opportunity of minimizing error rates, and thus ensuring customer satisfaction. Companies that have introduced this type of QM system generally also arrange for it to be certificated by an impartial organization. One organization in Germany carrying out such inspections is the Deutsche Gesellschaft zur Zertifizierung von QM-Systemen (DQS). Inspections are subject to regular repetition.

8.1.3 Total Quality Management System

The commercial situation in the world market and the increased demand for quality on the part of customers have set in motion learning and re-orientation processes in companies which are still far from completed. While quality assurance measures used to be largely aimed at finding and correcting problems, today the concept of quality in all areas of a company is gaining ascendancy. *Total quality management* (TQM) in the sense of general avoidance of problems and quality throughout the company is now the order of the day.

The evaluation system is an integrated approach based on internationally acknowledged criteria, e.g. in accordance with the model of the *European Foundation for Quality Management* (EFQM) [173]. Figure 8.1 clearly illustrates the change in the concept of quality from pure quality control of the product through the status of standard-oriented quality management to the all-encompassing understanding of quality.

Taking Siemens as an example, in the course of progressing from quality control to total quality management the company has combined product quality, process quality and company quality to form a strategic overall concept [174]. A full description and explanation of the QM system applied, however, would fall outside the scope of this book. The following subsections of Section 8 therefore aim to restrict themselves to product-specific and system-specific tests on high-voltage and extra high-voltage cables, including their accessories.

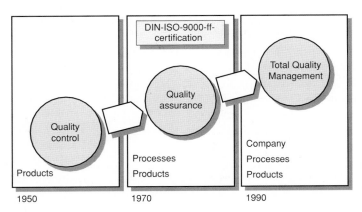

Figure 8.1
Development stages from quality control to total quality management (TQM)

8.2 Classification and Functions of Tests on Cables, Accessories and Systems

Tests can be classified according to different criteria, such as the standardization organizations which specify them (VDE, IEC, EdF . . .), the characteristics that are to be tested (dissipation factor, conductivity, electric strength), or the type of stress applied (AC voltage, DC voltage, impulse voltage) etc. In the case in point the tests, including the functions and aims allocated to them, should be tailored for the different stages in the development of a cable system, and should then also be used in the same order:

- *Development tests,*
- *Pre-qualification* and *type tests,*
- *Routine* and *sample tests* in the factory,
- *System tests* on site.

The details below are mainly concerned with conventionally-engineered cables, accessories and systems (i.e. systems with insulation made from impregnated paper or poly-

Table 8.1 Overview of tests on cables, accessories and cable systems

Type of test	Tasks / objectives	Time of test	Test piece
Development-tests	• Simulation of operating stresses • Estimation of long-term behavior by accelerated tests at increased loads • Determination of dimensioning-parameter values	• In phases of new and further-development	• Full-sized cables & accessories • Model cables • Test samples
Prequalification tests	• Proof of operational suitability • Estimation of long-term behavior by accelerated tests at increased loads • Confirmation of customer requirements	• At the start of important projects • If requested by customer	• Full-sized cables & accessories
Type tests	• Proof of operational suitability • Proof of properties specified in standards • Confirmation of customer requirements	• After conclusion-of development • On change of design and material • If requested by customer	• Full-sized cables & accessories
Sample-tests	• Proof of properties specified in standards	• After manufacture of specified part of order	• Individual constr. elements
Routine-tests	• Proof of properties specified in standards • Manufacturig check, quality assurance	• Before delivery to site	• Every length & accessory supplied*)
On-site tests	• Installation check, quality assurance • Proof of readiness to switch on • Estimation of ageing condition	• After laying • After repair • After long service time	• Complete cable systems

*) Routine tests only for prefabricated accessories (currently still at the consultation stage)

248

mers, particularly XLPE). The main focus here is on electrical behavior; tests on non-electrical properties are only staged in so far as they are necessary for an understanding of the wider context. Table 8.1 provides a preliminary overview of the tests classified in this way, their aims, when they are applied, and the preferred test objects to be used.

8.2.1 Development Tests

The nature and scope of tests to be carried out when developing high-voltage cables and accessories have not yet been specified through standardization, and it has therefore ultimately been left up to the individual producers to use their know-how and philosophy to design such tests [22]. However, in the early 1990s the Cigré internationally-based task force 21.03 worked out and published comprehensive recommendations for development tests on extra high-voltage cables with an extruded dielectric, including the associated accessories; in the meantime these recommendations have become sufficiently important to justify description as a 'quasi standard' [175].

Development tests can be carried out on so-called "full-sized" (scale 1:1) prototypes of the eventual product, on smaller scale models or just on separate parts (material samples). Experimental studies of this type have until now represented an important part of the overall development of cables and accessories, since the possible changes in characteristic over their expected service life of at least 40 years largely defy exact numerical calculations and computer simulations.

The main objectives of development tests therefore lie in 'fast-forward' estimation of the long-term behavior of cables and accessories that are to be installed in the network in future. Methods include artificially accelerated ageing under the influence of an appropriate degree of thermal/mechanical/electrical overstressing, along with creation of a statistically secure basis for the eventual dimensioning of the cables, as discussed in Section 3. Another function of these tests is to confirm through experiment the dependability of the dimensions of cables and accessories, which are initially calculated theoretically. This is approached by testing actual-size prototypes and taking into account all the operational stresses and overstresses that have to be expected in network operation.

As well as short-term tests to the point of failure, this also requires above all *long-term tests*, whose loading parameters (temperature, form of voltage, field strength, duration of test) are determined by the definitive failure mechanisms and the life cycle characteristics (see Section 3.2.2, Figure 3.12) of the dielectric used. This results in significant differences between development tests on cables with XLPE insulation, for example, and those with an impregnated paper dielectric [1]:

- *Impregnated-paper-insulated cables*
  As shown in Figure 4.10 (Section 4.1.3), the electric strength of paper-insulated cables achieves a static final value within a relatively short space of time, so that tests carried out over 20 days under thermal loading cycles are certainly meaningful enough to allow an estimation of the long-term behavior of the cables. The same applies to the accessories generally used with paper-insulated cables (Section 6.2.2).

---

[1] Developments with paper-insulated cables are now largely restricted to modifications for special customer requirements. For special factors in the context of polypropylene-coated paper see Section 3.4.2 and [34].

On the other hand, the relatively insignificant time dependence of the breakdown field strength of impregnated multi-layer dielectrics is responsible for the fact that the resistance to extreme short-term stress, e.g. impulse voltage, does not increase to the same extent as with extruded insulation. Impulse voltage tests to the point of failure therefore – after mechanical pre-stressing in the form of a *bending test* (see Section 8.2.2) – form an effective means at the new or adaptational development stage of testing out the electrical limits of paper-insulated cables and accessories and comparing them with the loads that can be expected under operating or test conditions [26]. At the same time, the statistical distribution of resistance to impulse voltage provides parameters that are important when applied to the question of cable dimensioning (Formula 3.6, Section 3.3.1). An example of relevant test results was presented above in Figure 4.9 (Section 4.1.3) in the form of two Weibull distributions.

Dielectric losses and the resulting heat at strong field that develops in cables with a conventional paper dielectric also plays a significant role (Section 3.4.2, Figure 3.26). Measurements of dissipation factor with the conductor at the max. permissible temperature and increased field strength (usually 1.7 times the operational stress) over a sufficient period of time (24 h) until a state of thermal balance is achieved accordingly lead one to expect important discoveries about the thermal reserves of paper-insulated high-voltage cables and accessories. To be absolutely certain that, even in the long term, no age-related deterioration of the tanδ will come into play, it is wise to check the dissipation factor regularly during the aforementioned long-term alternating voltage test taking place over 20 days.

Figure 8.2 provides a schematic representation of the patterns of dissipation factor over a period of time for thermally stable and unstable insulation: In both cases, once the nominal conductor temperature is reached at the beginning of the voltage loading a significant increase in dissipation factor is to be expected, based on the increased *conduction of ions* due to dielectric warming. In thermally stable insulation, however, a final static value must be established for the dissipation factor that represents a balance between heat generated by the dielectric or heat conducted into the dielectric from the conductor and the heat dissipated into the environment via the surface of the cable. On the other hand, if the loss that develops is too great, then the tanδ rises unrestricted, and the final result is *thermal breakdown* – a sign of thermally unstable insulation.

- *XLPE cables*
  In XLPE cables which leave the factory as thermally stable, low-loss cables, the dissipation factor and development of heat play a less important part. Instead, the time-

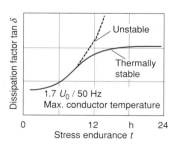

Figure 8.2

Development of dissipation factor in thermally stable and unstable paper cable insulation (schematic)

dependency of electrical strength (characteristic life curve as shown in Figure 3.13, Section 3.2.2) gains in significance. This factor is fed directly into the dimensioning of cable wall thickness via the so-called *ageing factor* k_t (Formula 3.14, Section 3.3.2), and also has a considerable influence on the statistical probability of failure in extended cable systems [21, 22]. The experimental determination of the characteristic life curve within the framework of long-term tests under alternating voltage stress thus represents a primary focus for development tests on XLPE cables [175].

Perfect "full-sized" cables, however, present the problem that breakdown tests do not generally produce any directly determinable results in the form of "active" break-downs; instead the result is in the format: "The strength is better than ..." (a particular limit value). An example of this type of pronouncement was presented above with the life diagram of high-voltage XLPE cables in Figure 4.32 (Section 4.2.2): with the exception of very few samples, all of which originated from the early days of manu-facturing XLPE cables at the beginning of the 1980s, none of the tests analyzed there resulted in a breakdown.

More detailed information on the long-term electrical behavior of high-quality XLPE dielectrics can therefore often only be obtained with the assistance of suitable *model tests*. This includes life tests carried out on samples of cable with identical insulation but considerably reduced thickness of around 6 mm, as is used in preference by Japan-ese manufacturers [176]. A further option for ascertaining the usable life characteris-tics of XLPE cable dielectrics through experimentation lies in using short-term voltage tests (so-called *step tests* or *ramp tests*) on *model cables* until the point of breakdown at the beginning and end of specified ageing periods. Such tests are carried out under the influence of increased field strengths, temperatures or temperature cycles and, if appropriate, other acceleration factors such as moisture (*residual strength measure-ments*) [21, 77]. Figure 8.3 shows for comparison two "characteristic life curves" that have been determined in this manner.

The primary aim of these long-term tests is to restrict the life exponent N as exactly as possible in the empirical life law $E^N \cdot t = $ const. (Formula 3.4, Section 3.2.2); this expo-nent is required for a number of statistical calculations in the development and dimen-sioning of the cable.

Breakdown tests in the step test with *impulse voltage* on real high-voltage XLPE cables likewise serve to determine statistically certain data for cable dimensioning. The most

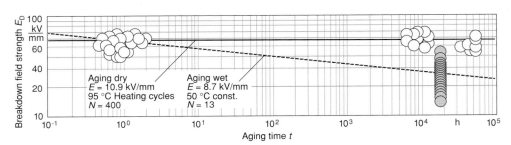

Figure 8.3
Analysis of ageing tests on XLPE model cables through measurement of residual strength at 50 Hz and representation in the life diagram [21]

important results these tests produce are the dimensioning parameters *Weibull slope b* (Formula 3.7) and *design impulse-field strength* E_{ds} (Formula 3.15). Further information on the impulse withstand voltage of XLPE insulation can be obtained by applying different temperatures and polarity sequences during the impulse voltage tests. Results from studies of this kind were presented earlier in the context of the electrical properties of high-voltage XLPE cables in Section 4.2.2. (Figures 4.34 to 4.36).

- *Accessories*

  The nature and scope of development tests carried out on accessories for XLPE cables is determined by the construction and insulation material (Section 6.2.3). A good starting point for choosing a material is to determine the most important electrical properties of the possible insulation materials. The first steps that need to be taken in this case are measurements of dissipation factor and breakdown tests on suitable samples. Figure 8.4 shows as an example the analyzed breakdown results of a silicone rubber with alternating voltage, as described in [131]. Amongst other things, these results substantiate the conclusion that the electric strength of this material is only affected very slightly by the duration that the voltage is applied and by the temperature.

  The most meaningful development tests on prototype cable accessories take place in conjunction with development of the cable itself, in that the correctly-installed sealing ends and joints are subjected to the same thermal and electrical stresses as the cables. As components of the same system later on in mains operation they will ultimately again be subject to the same stresses. By way of example Figure 8.5 shows the test setup for long-term alternating voltage tests on a 400 kV XLPE cable with outdoor sealing ends and two switchgear sealing ends connected opposite one another to a "back-to-back joint" (Section 6.2.3).

A brief overview of the development tests carried out on high-voltage and extra high-voltage cables and accessories, which are presented here as examples, is provided again in summary form in Table 8.2.

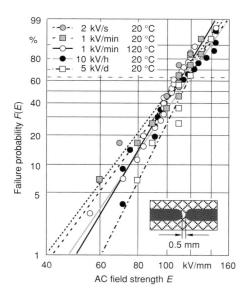

Figure 8.4
AC breakdown strength of silicone rubber with different test parameters (as described in [131])

Figure 8.5
Long-term AC voltage test on a 400 kV XLPE cable with accessories in the R&D center at the
Pirelli Kabel und Systeme cable factory in Berlin

Table 8.2
Development tests on cables and accessories for high voltage and extra-high voltage (extracts)

Object	Test
Paper-insulated cables incl. accessories	• Step-impulse voltage test (on cables after bending-test) • 20-d-alternating voltage test at 1.7 U_0 and heating cycles • Measurement of dissipation factor over 24 h at 1.7 U_0 and max. conductor temperature
XLPE cables	• Life tests under increased alternating voltages • Ageing test followed by measurement of residual strength • Step-impulse voltage test, where appropriate with max. conductor temperature and alternating polarities
Prefabricated accessories for XLPE cables	• Analysis of electrical and dielectric behavior of materials • Long-term AC voltage test under thermal load cycles • Impulse and AC voltage tests as for cables

8.2.2 Type Tests and Pre-Qualification Tests

Type tests and pre-qualification tests have a number of things in common, and for this
reason it was decided to deal with them together in this section.

- Both types of test are designed to gain the most complete picture possible of all the
  electrical, thermal, mechanical and chemical characteristics of the cables and accessories that are relevant to their intended application. Reliable conclusions can then be
  drawn concerning their dependability and future operational behavior in the power
  supply network.

- A prerequisite of the tests is that cables and accessories exist in a form ready for installation once their development is complete.

253

- The tests are carried out on short lengths of cable that is ready for mass production, including appropriate accessories where relevant.

- The test objects – possibly following on from additional visual inspections requiring some parts of the objects to be opened and dismantled – are destined for scrapping rather than operational use.

- The mechanical, thermal and electrical stresses can therefore be increased to the point where ageing processes and other irreversible pre-damage occurs in the dielectric.

- The tests are fully documented and – assuming a successful outcome – certified.

The differences between type tests and pre-qualification tests lie in the reason for the test, the test criteria and the testing location:

- *Type tests*
  Type tests are sometimes ordered by customers in connection with larger orders, but they are more commonly initiated by the cable manufacturers themselves, and carried out in their own laboratories. They frequently form the conclusion of a new development or an enhancement; the test criteria are defined directly in relevant standards (IEC, VDE etc.) or, if there are not yet any binding norms in existence for certain voltage classes or system components, then they are derived sensibly from existing standards for other technically similar equipment. Figure 8.6 shows a common setup for the type test on a high-voltage cable with two sealing ends and a joint, while Table 8.3 shows a summary of certain typical criteria for type tests on XLPE-insulated and paper-insulated high-voltage cables.

- *Pre-qualification tests*
  Unlike the standardized type tests, pre-qualification tests are normally performed on customers' instructions; the customers also stipulate the test criteria and – unless they have a suitably-equipped laboratory themselves – commission an impartial organization to carry out the tests. Pre-qualification tests are often directly related to planned sizable new installations using leading-edge technology, for which there is consequently either no or only limited operational experience. A general announcement is issued about a pre-qualification test to enable several cable manufacturers to take part

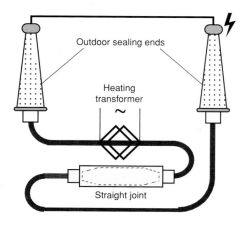

Figure 8.6
Layout for type test on a high voltage cable
with sealing ends and straight joint

Table 8.3
Characteristic components of type tests on XLPE- and paper-insulated
high voltage cables (extracts)

Test	Applied to...
Bending test	• Paper-insulated cables • Polymer-insulated cables
Thermal load cycles under alternating voltage	• Paper-insulated cables • Polymer-insulated cables • Accessories[1]
Measurement of dissipation factor at max. conductor temperature	• Paper-insulated cables • Polymer-insulated cables
Impulse voltage test at 5–10 °C above max. conductor temp.	• Paper-insulated cables • Polymer-insulated cables • Accessories[1]
Measurement of partial discharges	• Polymer-insulated cables • Prefabricated accessories[1]
Measurement of conductive layer resistance	• Polymer-insulated cables
Corrosion protection test with impulse voltage and direct voltage	• Cable sheaths • Joint housings[1]
Testing of separating points with impulse voltage and direct voltage	• Separating joints for "cross bonding"

[1] Tests on accessories when specified, otherwise by agreement

at the same time. In this way the leading operator, who also specifies the criteria, can gain a comprehensive overview of the current state of the art in the chosen technical field; at the same time they gain an impression of the know-how of the cable manufacturers who are involved, and this helps with decision making later at the time of placing an order. The duration and requirements of a pre-qualification test usually exceed those of the comparable type tests by a considerable margin; the type test often forms a requirement for inclusion in the pre-qualification test, or it is repeated as part of the latter test.

Important pre-qualification tests that have been organized in the fairly recent past have concentrated on the evaluation of XLPE cables and accessories for 400 to 500 kV, and were conducted under the leadership of Bewag in Berlin [129] and several large power supply companies from Japan [79]. In France, on the instigation of the EdF, pre-qualification tests on high-voltage cables and accessories have even been incorporated as an established part of the national system of standardization (see Section 8.4.2). As an example of the structure and requirements of this type of long-term test, Figure 8.7 shows the arrangement specified by Bewag for its second pre-qualification test on 400 kV XLPE cables and accessories at CESI in Milan in 1996, with six European cable manufacturers taking part; the arrangement is based on the Cigré recommendations in [175].

There is also a variation on the procedure described above, whereby power supply companies generally demand a one-off pre-qualification before accepting a manufacturer as a potential supplier of cables of a particular construction and for specified

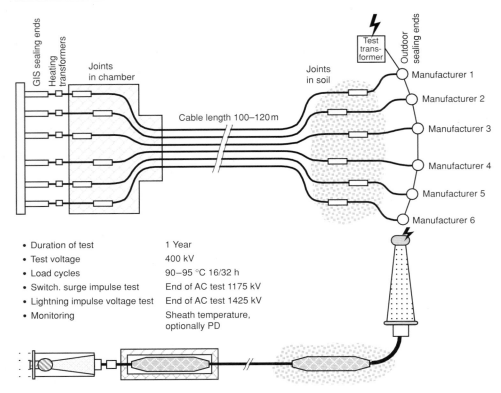

- Duration of test — 1 Year
- Test voltage — 400 kV
- Load cycles — 90–95 °C 16/32 h
- Switch. surge impulse test — End of AC test 1175 kV
- Lightning impulse voltage test — End of AC test 1425 kV
- Monitoring — Sheath temperature, optionally PD

Figure 8.7
Layout and main requirements of a prequalification test on 400 kV XLPE cables and accessories (as described by Bewag/Berlin [129])

voltage classes. Examples of this are the French national power supply company Electricité de France (EdF) and, although relating to the field of XLPE-insulated medium voltage cables, most of the German cable users, who demand an ageing test running for two years as pre-qualification [177].

8.2.3 Routine Tests and Sample Tests

Routine/unit tests and sample tests are generally closely linked with customer orders, and are carried out by the manufacturer on items currently in production. The criteria, which are specifically geared for the type and voltage class of the cables or accessories to be tested (where these are prefabricated models), are specified in norms and standards and must be seen as binding; their primary function is quality assurance.

- *Sample tests*
  Sample tests (*special tests*) are generally only applicable to cables, and fall due with large orders after specified lengths (several km) have been manufactured. They include parts of a type test (e.g. step/impulse voltage tests at max. conductor temperature) which cannot be carried out on the whole shipment for technical reasons, or which would lead to a danger of lasting damage [178].

- *Unit tests* or *routine tests*

  These tests, on the other hand, must be carried out on every length of cable supplied and – if specified – on every (prefabricated, and therefore factory-testable) accessory; they always include a *short-term alternating voltage test* at ambient temperature, irrespective of the type of dielectric.

Since the equipment must be able to withstand the continuous stress of network operation for at least 40 years following installation of the cable system, great care must be exercised when choosing the voltage level in the course of the routine test to prevent any premature ageing or other preliminary damage from being initiated. The aim is therefore to arrive at a balanced compromise between formulating a test that is meaningful and ensuring that it will not prove to be damaging. This is particularly relevant for extra high-voltage cables, which are subject to relatively high electrical stresses even under operating conditions, and must therefore not be overstressed during the test.

With reference to PE/XLPE cables the Cigré Working Group 21.03 recommended an upper limit of 27 kV/mm for the test field strength, and this should never be exceeded in routine tests [178]. Instead *partial discharge measurement* has proved reliable over many years at detecting errors in extruded dielectrics (see Section 8.3). Rather than PD measurement, for cables with impregnated paper insulation with their proven effective resistance to partial discharges the *measurement of dissipation factor* in every section supplied forms part of the scope of the routine tests.

The most important electrical tests that form part of routine testing on paper-insulated and polymer-insulated high-voltage cables and prefabricated field control elements for accessories for polymer-insulated cables are compiled once more in Table 8.4.

Routine tests on high-voltage and extra high-voltage cables, with the costs involved in setting them up for each cable section or accessory supplied, add a quite considerable cost factor. In consequence of this fact the cable manufacturers, to some extent in cooperation with the suppliers of their test equipment, have developed a number of solutions for rationalizing the preparatory work for the routine tests. Most importantly, these solutions include sufficiently surge-proof and possibly also partial-discharge-free sealing ends for short-term alternating voltage tests and PD measurements on cables. Two different techniques have become established for this task according to the type of cable insulation:

Table 8.4 Key components of the routine test

Test	Applied to...
Short-term alternating voltage test	• Paper-insulated cables • Polymer-insulated cables • Prefabricated accessories[1]
Measurement of dissipation factor	• Paper-insulated cables
PD measurement	• Polymer-insulated cables • Prefabricated accessories[1]

[1] Tests on accessories where specified, otherwise by agreement

- Prefabricated *stress cones* on laminated-paper tubes, using the principle of capacitor stress controlling as illustrated in Figure 6.5 (Section 6.1.2) for voltage testing of paper-insulated cables and

- *Water sealing ends* for the voltage testing and PD testing of polymer-insulated cables. A field control principle which is otherwise only used in the mains network for medium voltage cable sealing ends is here transferred to short-term operation under extra high-voltage stress: the so-called *refractive-resistive field control* [8, p. 426]. It takes advantage of the high permittivity in conjunction with the comparatively low ohmic resistance of the water in a double-walled insulating tube in order to accurately influence the field distribution at the cable end. In the process the potential gradient along the cable end is comparatively modified by the automatic control of the water temperature and resistance in the circulation process to the point at which no creepage discharges or even flashovers occur [179].

With these sealing end systems the work involved in so-called *end preparation* on high-voltage and extra high-voltage cables that are awaiting voltage tests is essentially reduced to stripping the cable sheath and wire screen (if there is one) and the outer field-smoothing layer over a length of the order of 1 to 3 m depending on the level of the voltage test. In the case of paper-insulated cables, after the condensor cones have been mounted their ends must still be sealed by winding so that the remaining gap between the paper insulation and the stress cone's support pipe can be filled with cable oil; on the other hand the water sealing ends for polymer-insulated cables are sealed using silicone plugs matched to the diameter of the conductor. As an example of this type of test setup, Figure 8.8 shows a 400 kV XLPE cable with water sealing ends which are designed for operation with up to 600 kV.

Similar systems are used for routine tests on prefabricated field control elements for polymer-insulated cable sealing ends. As shown in Figure 8.9 they consist of a short sec-

Figure 8.8
AC voltage test on a 400 kV XLPE cable using water sealing ends (Pirelli Kabel und Systeme cable factory, Berlin)

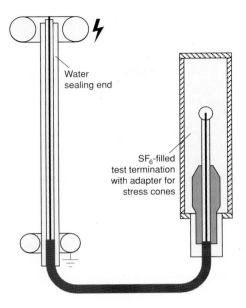

Figure 8.9
Voltage test on prefabricated stress cones
using a water sealing end and universal test
adapter in SF$_6$

tion of cable with a water sealing end on the one side, and the adapter for accommodating the stress cones in an SF$_6$-filled test termination on the other side.

In comparison, voltage tests on prefabricated joints turn out to be a great deal more demanding of resources. The body of the joint, even if it is already in use by pre-prepared sections of cable, must be almost fully assembled and provided with a grounded sheath. Only potential control at the open cable ends succeeds in reducing the installation overhead even here by means of water sealing ends.

8.2.4 Tests on Cable Systems

Tests on cable systems, otherwise known by the term *on-site tests*, represent the last step before commissioning newly-installed systems. These tests therefore have the task of detecting faults that may have arisen during transportation, laying and installation of accessories. In the case of accessories produced on site (e.g. lapped sealing ends and joints, and extrusion molded joints – Sections 6.2.2 and 6.2.3) the on-site test simultaneously takes on the role of a routine test, as already performed on prefabricated accessories in the factory. The same task presents itself again after cables have been repaired, for example, by installing new straight joints. Additionally, a number of cable system operators use on-site tests at regular intervals in order to estimate the state of play in terms of ageing, and possibly to detect faults in advance as they begin to develop (so-called *repeat tests*).

A distinction should always be made between two different types of system test:

- Testing of the internal insulation, or the insulation capability, which is comparable to the voltage test in the factory, and

- Testing of the cable sheaths and corrosion protection. The tests are designed to provide information on whether the non-metallic coverings of the cables are in perfect condition, and whether faulty seals or corrosion damage are likely to develop on metal sheaths.

Corrosion protection test

The *sheath tests* and *corrosion protection tests* are performed on high-voltage cables of all constructions using a moderately high DC voltage (1 kV in the case of gas pressure cables with welded external pipes, otherwise 5 to10 kV) for the duration of one minute between metal sheath or screen and ground [149, 180]. The criterion for the undamaged condition of the sheath is its insulation resistance, which should be at least 2.5 MΩ per km length of cable. Values lower than that indicate the presence of damage to the sheath or (passive) corrosion protection[1], which must then be located using suitable methods [8, p. 452 ff.].

Reliable detection of damage depends on a sufficiently conductive "counter electrode", so that when a fault current occurs it can be conducted back to the source. For this purpose, the cables must at least be fully covered with sand and it may be necessary to dampen the sand before the test. The return line can also be improved by graphiting the outside of the cable sheaths. Needless to say, before the DC voltage test is applied to the corrosion protection, all the ground connections of the metal sheath or screen and of the accessory housing connected to it must be disconnected.

Some manufacturers are of the opinion that as long as the cable passed the routine test in the factory, the corrosion protection test after laying is at the same time sufficient to confirm that the dielectric is intact, and that there is therefore no need to test the insulation on site.

Testing the insulation

Testing the (internal) insulation of cables and accessories proves to be considerably more demanding of resources. In order to make sure that the test is accurate enough, of course, the stress applied must where possible exceed the operational load by a considerable margin. The form of voltage suitable for this purpose depends on the type of dielectric to be tested.

- *Impregnated paper dielectric*
  In the case of impregnated paper insulation the use of DC voltage has proved effective over a period of many decades, and relatively low-output generators are sufficient to produce such voltage on site even for lengthy runs of cable because of the negligibly low charging load. However, because of the absence of polarity changing, DC voltage must be set several times higher than would be the case if an AC test voltage were applied. Depending on the size of the rated voltage, the relevant standards recommend a DC voltage test level between 3 U_0 ($U_N \geq 400$ kV) and 4 to 4.5 U_0 ($U_N = 110$ kV) [181]. This means that for on-site testing of paper-insulated 400 kV cables ($U_0 = 230$ kV) with DC voltage, a generator producing at least in the region of 700 kV is required, and it must also be a fairly large one (see Figure 8.10).

---

[1] A resistance measurement with just 50 V DC is sufficient for testing active (cathodic) corrosion protection that may be in place [149].

Figure 8.10
DC voltage generator for the on-site testing of 400 kV oil-filled paper-insulated cables

- *Extruded dielectric*
  On the other hand, experts nearly unanimously shun the DC voltage test for polymer-insulated cables, and especially those with a PE/XLPE dielectric (e.g. [128, 180-184]). In combination with high-voltage cables and extra high-voltage cables with their laterally watertight construction (Section 2.3.4, Table 2.4), the danger of preliminary damage due to the application of DC voltage [185], which was discussed intensively for water-tree-aged cables, plays a less important part than the discovery that DC voltage is not suitable for detecting faults in polymer-insulated cables and accessories. Thus the view was put forward as long ago as in 1982, at the Cigré conference in Paris, that the DC voltage test does not represent a usable method of detecting faults in cables with PE/XLPE insulation [182].

The definitive reason for this has been recognized as the lack of polarity changing and the resulting space charging effects under the influence of DC voltage [185]. A number of proposals have emerged from the search for suitable alternatives, recommending various forms of voltages that can be changed with respect to time. However, the majority of these proposals have been aimed primarily at the testing of medium voltage cables [128, 186]. For the on-site testing of high-voltage and extra high-voltage cables only two of these alternatives have proved suitable. These are the only methods that can be reproduced on a construction site without the need for an unjustifiable outlay of technical resources.

- *Oscillating waves*, a type of damped-oscillating switching voltage illustrated in Figure 8.11, which was successfully trialed by the Cigré task force 21-09 in a succes-

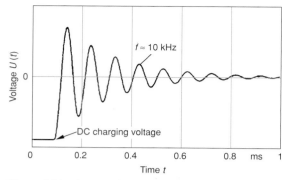

Figure 8.11 Pattern of "oscillating waves" by time for on-site testing

Figure 8.12 Mobile resonance testing system for 160 kV / 40 A

sion of prototype tests and was subsequently presented to experts in the field [187] and

- Sinusoidal alternating voltage with a frequency approaching that in the network which, in simplified terms, is generated in a *resonant circuit* with the cable to be tested as capacitance and a movable reactor as an oscillating circuit inductor [188, 189]. The reactor, frequency-controlled power transformer and all the necessary supplementary equipment required are installed on a low loader, and they can thus be transported easily to the various locations in which they are to be used (Figure 8.12). The resonance balancing of the test system on cables of differing lengths and conductor cross sections (and thus varying capacitance) is achieved by switching over the reactor and varying the frequency between 35 and 70 Hz. Basic studies have established that the breakdown behavior of the insulation within this frequency range is not significantly different from the behavior at 50 Hz [190]. This system thus makes it possible to generate the same test loading as is applied for routine testing in the factory.

8.3 Partial Discharge Tests on Polymer-Insulated Cables

8.3.1 Basic Principles

A partial discharge occurs in polymer dielectrics if the electric strength is exceeded at a locally restricted point, and if a local breakdown occurs there as a result (Section 3.2.1, Figure 3.9). The best way to explain the processes taking place here is by using the well-known simplified equivalent circuit diagram comprising three capacitances representing the void itself, the dielectric connected in series with it and the intact dielectric connected in parallel with them both (Figure 8.13). Parallel to the void the equivalent circuit diagram provides a spark gap which breaks down when a specific voltage U_z, assumed to be constant, is exceeded, and this has the effect of discharging the capacitor. This results in:

- On the one hand repeated voltage collapses at capacitor $C_{1,}$, as indicated in the oscillogram in Figure 8.14, and

- On the other hand pulse-formed equalizing currents within the capacitor arrangement and its incoming lines, by means of which the charge (or energy $1/2 \cdot C_1 U_1^2$) that has

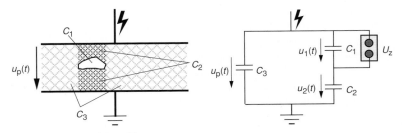

U_z: Ignition voltage of the void
C_1: Capacitance of the void
C_2: Capacitance of the health dielectric in series with the void
C_3: Capacitance of the health dielectric parallel to the void

Figure 8.13
Simplified equivalent circuit diagram to represent a void with partial discharges in polymer-insulated cables

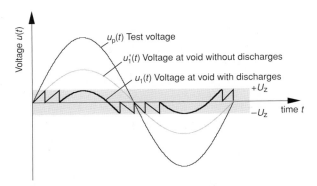

Figure 8.14
Voltage patterns in the equivalent circuit diagram 8.13 when C_1 is repeatedly discharged ("PD oscillogram")

263

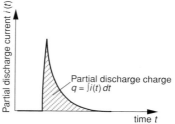

Partial discharge charge
$q = \int i(t)\, dt$

Figure 8.15 Partial discharge pulse and PD charge

been converted to heat at the "spark gap" is supplied. Only these high-frequency current pulses and the current-time area that they delineate (*PD charge* as shown in Figure 8.15) can be evaluated for partial discharge measurement.

The recharging pulses that actually occur when C_1 breaks down at the location of the fault are, however, quite different from those which can be measured externally, i.e. at the cable end. Part of the charge is supplied initially from the "parallel capacitance" C_3, i.e. from the cable itself and, in addition, the shape of the pulse is distorted to a greater or lesser extent on the way from the fault location to the measuring device depending on the type and length of cable. The PD charge that can be measured externally thus represents only the *apparent charge*, which is generally lower than the actual charge at the location of the fault. Ultimately all the other elements contained in the test circuit, in particular the transmission characteristics of the coupling unit and the measuring leads, and the evaluation in the *PD detector* itself play a crucial part in what is shown at the output as the PD charge q. For this reason a PD test on cables generally comprises three steps:

- Partial-discharge-free construction of the test circuit and installation of the measuring system;

- *Calibration* of the entire circuit including test object with calibration pulses fed in from outside at low voltage, so that there is no possibility that any internal disturbing pulses can occur yet (battery-driven pulse generators are commonly used, or otherwise pulse generators driven by mains voltage no higher than 220–240 V);

- The actual measurement of PD at the stipulated test voltage. This is basically a comparison between the signals being received by the detector from the test object and those stored previously from the calibration generator.

For PD testing on XLPE high-voltage and extra high-voltage cables the relevant regulations generally require a calibration with 5 pC that is detected reliably, i.e. must exceed any *noise level* that may be present by at least 100%. Figure 8.16 illustrates the relationships using the oscillogram of an analog amplifier in the time range. Usually, in the same way, pulses are represented in *Lissajou's figures*, which recreate a 50-Hz period as an ellipse, and increasingly also digital displays. The actual partial discharge tests represent an important quality assurance measure, and must be carried out at voltages between 1.5 U_0 and 2 U_0 on every section of cable supplied, and generally also on prefabricated accessories. The PD charge q that can be detected by this method, also specified as *PD intensity,* must under no circumstances exceed 5 pC; in the meantime invitations to tender, at least from German cable operators, often stipulate verifiable limits ≤ 1 pC – and correspondingly a calibration at 2 pC.

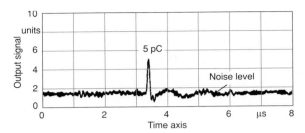

Figure 8.16 Calibrating an analog PD measurement system at 5 pC

8.3.2 Acceptance Test in the Factory

In order to reliably achieve the verifiable limit for partial discharges with an order of magnitude of 1 pC even for longer sections, the leading manufacturers of polymer-insulated cables have largely equipped themselves with fully-electromagnetically-shielded test laboratories including effective high-frequency filters in all the electrical supply lines in order to exclude all kinds of external interference (see Section 7.4.3). The crucial value for assessing the electromagnetic shielding is the so-called *shield attenuation* in the relevant frequency range for PD tests. The shield attenuation a indicates in logarithmic form the relationship $\hat{A}/\hat{A}_0$ between the amplitudes of an electromagnetic signal outside and inside the shielded area, and is measured in dB:

$$a = 20 \cdot \lg(\hat{A}/\hat{A}_0) \text{ in dB} \tag{8.1}$$

To illustrate by example the shield attenuations that can be achieved today, Figure 8.17 shows the attenuation pattern of the 600 kV test laboratory commissioned by the Siemens cable factory in Berlin in 1993. Allowance should be made here for the fact that the curve throughout the frequency range 10 kHz and 10 MHz marks the upper measuring limit, and thus the actual shield attenuation is in each case higher than that shown, at approx. 80 to just under 120 dB. When the values are recalculated in accordance with Formula 8.1, this result means that the amplitudes of external interference signals are damped by the laboratory walls and line filters by factors of at least 10^4 at 10 kHz (80 dB) to almost 10^6 above 1 MHz (just under 120 dB).

Partial discharge measurement

The test object (capacitance C_x) is connected to the high-voltage transformer as shown in Figure 8.18 *in parallel* with a *coupling capacitor* C_k and through an incoming line

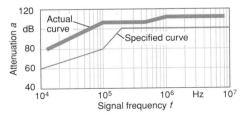

Figure 8.17 Shield attenuation of modern PD test laboratories (Siemens cable factory, Berlin)

265

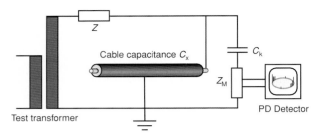

Figure 8.18   Circuit for PD testing on cables

impedance Z as a harmonic filter. The measurement signal can be captured at the *quadruple coupling circuit* Z_M that is connected in series with C_k with a suitable R-L-C circuit, and transferred to the *PD detector* via a coaxial line. In the electromagnetic-interference-free environment of a shielded laboratory the quadruple circuit and detector are usually laid out as a *wide-band system* [15, p. 334 ff.], and, in the event that partial discharges occur in the test circuit, provide the *apparent PD charge* as defined in Section 8.3.1 as an output signal.

In the now fully-installed state of the test circuit, the entire system is calibrated as the first step in PD measurement. The calibration is performed by feeding in a partial discharge pulse that has been generated at low voltage and has a precisely defined charge (usually 5 pC), then storing and logging the output signal from the detector. In the case of PD tests on long cables, it is sometimes useful to perform the calibration operation from both ends of the cable. Following this, the AC test voltage is set to the level as specified in the relevant standard or as agreed between the manufacturer and the client, maintained if applicable at a constant level for a specified length of time before the output signal from the PD detector is captured. If this signal is anything other than "zero", then the PD strength q must be determined in pC by comparing the response from the detector with the calibration pulse, and then logged. The test can be regarded as successfully concluded if q remains below the specified maximum value.

In faulty cables, on the other hand, an output signal q occurs at the detector, and its voltage dependence is according to figure 8.19 characterized by three ranges:

- Up to the so-called *PD inception voltage* U_e no discharges, $q = 0$;

- Above U_e, increase of PD charge that initially is almost in proportion with the voltage, $q \sim U$;

- A limit value is finally reached, $q \approx \text{const.}$

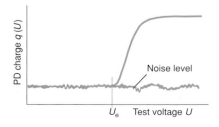

Figure 8.19
The relationship between voltage and the PD charge of faulty polymer-insulated cables (schematic representation)

8.4.2 Selected Standards for High-Voltage and Extra High-Voltage Cables

On both the European and national levels there are just four standards corresponding to the cables of conventional design used today in high-voltage and extra high-voltage systems: low-pressure oil-filled cable, internal and external gas pressure cable and polymer-insulated cable. On an international level (IEC, Table 8.6) these are supplemented by an additional specification for high-pressure oil-filled cables (Figure 4.18), which are less commonly used in Europe now. All the standards apply to both cables and accessories (for polymer-insulated cable some of the specifications are still in preparation) and, in accordance with the details in the preceding section, they specify only the test requirements. Therefore, with the exception of the European and German editions of the regulations for polymer-insulated cable they are consistently overwritten with "Tests on...". Table 8.6 contains an overview of the current standards for high-voltage and extra high-voltage cable; Tables 8.7 and 8.8 list the titles of the IEC regulations and the DIN VDE regulations respectively.

Table 8.6
Standards for high voltage cables and accessories at international, European and German national levels [192]

Application area	IEC Publications	Harmoniz'n documents	Previous national regulations	Future national regulations: DIN VDE...
LPOF cable	IEC 60141-1[1]	HD 633	DIN VDE 0256	0276-633 Parts 1, 2, 3D
Int. gas press. cable	IEC 60141-2[1]	HD 634	DIN VDE 0258	0276-634 Parts 1, 2, 3C
Ext. gas press. cable	IEC 60141-3[1]	HD 635	DIN VDE 0257	0276-635 Parts 1, 2, 3C
HPOF cable	IEC 60141-4[1]	–	–	–
Polymer-ins. cable	IEC 60840	HD 632	DIN VDE 0263	0276-632 Parts 1, 2, 3D, 4D, 5D

[1] Valid only in connection with IEC 60141 (see Tab. 8.7)

Table 8.7 Titles of IEC publications for high-voltage and extra high-voltage cables

IEC Publ. no..	Title
60141	Tests on oil-filled and gas-pressure cables and accessories
60141-1 (1993)	Part 1: Oil-filled, paper-insulated, metal-sheathed cables and accessories for alternating voltages up to and including 400 kV. Amendment 1 (1995)
60141-2 (1963)	Part 2: Internal gas-pressure cables and accessories for alternating voltages up to 275 kV. Amendment No. 1 (1967)
60141-3 (1963)	Part 3: External gas-pressure (gas compression) cables and accessories for alternating voltages up to 275 kV. Amendment No. 1 (1967)
60141-4 (1980)	Part 4: Oil-impregnated paper-insulated high-pressure oil-filled pipe-type cables and accessories for alternating voltages up to and including 400 kV. Amendment No.1 (1990)
60840 (1988)[1]	Tests for power cables with extruded insulation for rated voltages above 30 kV (U_m = 36 kV) up to 150 kV (U_m = 170 kV). Amendment 2 (1993)

[1] New edition also for accessories 1998; separate publication in preparation for voltages up to 500 kV

Table 8.8

Titles of the DIN-VDE regulations for high voltage and extra high voltage cables

DIN VDE no.	Title
0276-632 Parts 1, 2, 3D, 4D, 5D	Cables with extruded insulation and their accessories for nominal voltages above 36 kV (U_m = 42 kV) to 150 kV (U_m = 170 kV)
0276-633 Parts 1, 2, 3D	Tests on oil-filled (fluid-filled) paper- and PPL-insulated cables with metal sheaths and their accessories up to and incl. 400 kV (U_m = 420 kV)
0276-634 Parts 1, 2, 3C	Tests on internal gas pressure cables and their accessories for alternating voltages up to 275 kV (U_m = 300 kV)
0276-635 Parts 1, 2, 3C	Tests on external gas pressure cables and their accessories for alternating voltages up to 275 kV (U_m = 300 kV)

Apart from the regulations listed here, which are formulated specifically for high-voltage and extra high-voltage cables and their accessories, when they are applied a large number of other specifications come into play (so-called *reference standards*); these are referred to in detail in the individual regulations. They concern, for example, the wave forms of the voltages that are to be used for various tests, the precision of measuring instruments, parameters for analyzing tests results etc. However, it is beyond the scope of this work to attempt to list all the reference standards.

Instead, in connection with test requirements for high-voltage cables and accessories with polymer insulation, it suffices to finish by mentioning once again the recommendations produced on behalf of the Cigré Study Committee SC 21 [175, 178] as well as the French national standard C33-253 [193].

The test recommendations of Cigré-SC-21, with the collaboration of many experts from around the world, have produced a scientific basis on which, from now on, the various standardization organizations can draw up regulations that may still be needed. At the same time this research equips operators for formulating sensible requirements for technically meaningful pre-qualification tests on polymer-insulated extra high-voltage cables.

The French standard C33-253 is noteworthy because of the fact that it has already integrated a pre-qualification test similar to that described in [129] as a fixed and compulsory part of the regulation, and thus applies to all suppliers. This standard also differs from the international and European regulations for high-voltage cable described above in that it indirectly stipulates certain constructional requirements concerning the thickness of the insulating wall by specifying compulsory maximum field strengths at the inner and outer conductive layers of the polymer dielectric. So a little more time and extensive harmonization work will be required before the cable market in Europe becomes a truly "common market" with uniform norms and standards.

9 Development Trends

In the period from the mid 1960s to the end of the 1970s, the cable industry and the research institutes concentrated on developing cables and cable systems for transmitting ever more power at ever higher voltages. As a result of the oil crisis and consequently slower growth in energy consumption and with energy prices rising rapidly at the same time, development priorities changed. On the one hand, development effort was directed at a number of spectacular projects for bulk-power transmission (see Table 5.12) while, on the other hand, after the mid 1980s increasing system capacity was no longer the top priority, at least in terms of medium-term targets. The focus was now instead on reducing investment and operating costs and improving the reliability and environmental compatibility of the existing cable types. Above all, great efforts were made around the world to replace paper/oil dielectrics with XLPE dielectrics, even in the high-voltage and extra high-voltage fields.

With the discovery of the high-temperature superconductor and the search for an alternative to extra high-voltage overhead lines that was made necessary for environmental reasons, the range of products for bulk-power underground transmission was extended again in the 1990s with the advent of gas-insulated lines (GIL) and HTSC cables.

No dramatic changes have recently occured in the engineering of high-voltage or extra high-voltage cables [194]. Most operators are extremely cautious when it comes to changing to new types of design, new dielectrics or new areas of technology. The reasons behind this are the well-known reliability of the paper-insulated oil-filled cables and the service life of this type of cable, which has been proven in practice to be over 40 years, two qualities that operators are anxious to retain.

However, for environmental reasons in particular, the life cycle of oil-impregnated paper-insulated cable is now coming to an end. XLPE-insulated cables are clearly winning an increasing market share even in the high-voltage and extra high-voltage field, and their technical qualities, in particular reliability and safety in operation, have been improved time and time again. The trend towards XLPE is clearly visible! When the first longer 400 kV XLPE cable systems are commissioned in 1998 for supplying electricity to conurbations in Europe the polymer-insulated cable will make the final breakthrough and will gradually squeeze out the paper-insulated oil-filled cable for all voltages up to 500 kV. Despite this, it is to be expected that paper-insulated oil-filled cables will continue to be ordered, manufactured and supplied well beyond the year 2000.

With the availability of extruded-insulated cables right through the range of voltages up to 500 kV, underground power transmission in cables has made a major step forward. Figure 9.1 illustrates the life cycle of technical progress with the paper-insulated oil-filled cable and the XLPE cable. The advances that can be made by using the extruded dielectric are mainly connected with environmental compatibility and operating costs.

The only major remaining problems for developing high-voltage and extra high-voltage cables further now lie in XLPE cables, and the engineering of the related accessories and systems. The fundamental problems and goals are compiled in Table 9.1. Most of the tasks have already been discussed in the preceding sections. Other details can be found

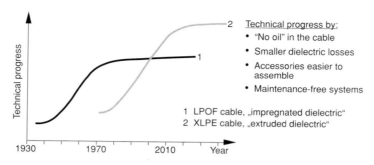

Figure 9.1 Lifecycle of high voltage and extra high voltage cables

in the numerous publications of Cigré study committee 21 [195–199] and other specialist articles [e.g. 200–202].

Even paper-insulated cables (low-pressure oil-filled and gas pressure cables in steel pipes), which number amongst the fully-developed means of transmission, still offer a certain amount of scope for technical enhancements. Examples are the use of polypropylene-paper laminates for oil-filled cables up to 500 kV (765 kV) and the planned retrofitting of 110 kV gas pressure cables by XLPE cables in steel pipe [203].

Table 9.1 Further developments in high voltage and extra high voltage cable systems

Field	Problems	Objectives	Market interest
LPOF cables	• Environmental compatibility • Dielectric losses at high voltages	• Impregnation medium not water-polluting • Replace paper with PPLP	↘
Cable in steel pipe	• Load capacity	• Retrofitting with XLPE, poss. HTSC	→
Polymer insulated cables	• Thick insulation • Operating experience with long lengths • Need for polymer cross-linking	• Increased operating field strengths • Use at U ≥ 400 kV • Non-cross-linked dielectric	↗
Accessories for polymer insulated cables	• Multiplicity of types • Filling material in sealing ends • Fitting costs & safety • Testing after fitting	• Modular design • "Dry" sealing ends • Prefabricated, tested accessories of slip-on and plug-in design • Selective PD measurement on site	↗
System engineering	• Cable laying techniques • AC testing before commissioning • Maintenance costs • Preventive maintenance • EMC	• Laying in tunnels; boring and ploughing techniques • AC resonance tests on site • Maintenance-free systems • Monitoring and diagnosis • Optimized laying arrangement; compensating conductor	↗

Table 9.2

Development trends in DC and three-phase AC bulk-power transmission systems

Field	Problems	Objectives	Market interest
Force-cooled conventional cables	• Cooling of cables laid in tunnels • Maintenance • Cooling of polymer insul. cables	• Air cooling • Cost reduction • Optimization of accessory cooling	→
HVDC cables	• Voltage limitations • Power limitations • Impregnated dielectric • Alternative to OH lines over long distances	• Increase of rated voltage >450 kV • Increase of power >600 MW • Introduction of extruded dielectrics • Development of HTSC cables as HVDC overland cable	↗
Gas-insulated lines (GIL)	• Environmental compatibility • Laying methods • Alternative to OH lines over long distances	• Replacement for SF_6 insulating gas • Optimization of laying in tunnel; pipeline technology for laying in ground • Cost reduction	Still limited
High-temperature superconductor cables (HTSC)	• Technical feasability; Economic viability • Dielectric • Cost of cooling	• Evidence by 2005 (prototype system) • Use of extruded dielectrics • HTSC materials with higher critical temperature	Still limited

The development trends in three-phase AC and DC bulk-power transmission systems are summarized in Table 9.2. The technical specifications of these systems were covered in detail in Sections 4 and 5 above.

High-voltage DC cables (in submarine and underground versions) will need to be developed for higher transmission capacities per system, and thus also for higher voltages, in order to match better the opportunities provided by modern converters. The replacement of the paper dielectric by an extruded dielectric in the HVDC cable is another objective that is extremely demanding from a technical point of view.

The transition in high-voltage cables from copper conductors to high-temperature superconductors, which now looks likely to be achievable some time between 2005 and 2010, is also likely to be carried out as a 'rolling replacement'. However, if superconductors with a critical temperature around the ambient temperature were to be discovered a leap forward in technology would result: 300 K superconductors would then revolutionize the whole field of electrical engineering. This could, for example, mean that cables would no longer require cooling, and underground power transmission could conceivably achieve extremely high power levels at voltages of just a few kV. This vision would signify the transition from high-voltage to high-current technology. However, this development is not likely to come in the near future.

10 Bibliography

[1]   N. N.: Die öffentliche Elektrizitätsversorgung 1995. Brochure from the Vereini-
       gung Deutscher Elektrizitätswerke VDEW e.V., Frankfurt/Main, Sept. 1996

[2]   Haubrich, H.-J.: Technisch-wirtschaftliche Aspekte der Integration von Kabeln in
       110 kV-Freileitungsnetze. Elektrizitätswirtschaft, year. 85 (1986) H. 21

[3]   Fricke, K.-G.; Paschen, R.; Steckel, R.-D.: AC Overhead and Underground HV
       Lines – Comparison and New Aspects. Cigré Session, Paris 1996, Contribution
       21–107

[4]   McMahon, M. et al.: Comparison of overhead lines and underground cables for
       electricity transmission. Cigré Session, Paris 1996, Contribution 21/22–01

[5]   N. N.: Calculation of the Continuous Current Rating of Cables (100 % Load Fac-
       tor). IEC Recomm. Publ. No. 287, 1969

[6]   Metra, P.; Rühe, L.: Milliken Conductors as Compared with Shaped Wire Hollow
       Conductors for Single-Core Oil-Filled Cables. Siemens R & D report Vol. 12
       (1983) No. 4

[7]   Takaoka, M.; Ono, M.; Kaji, I.: Development of 275 kV XLPE cable system and
       prospect of 500 kV XLPE cable. IEEE Trans. on Power Appararus & Systems
       Vol. PAS-102 (1983) S. 3254–3264

[8]   Heinhold, L. (Herausgeber): Kabel und Leitungen für Starkstrom, Part 1. Verlag
       Siemens AG Berlin &. Munich, 4. 1987 edition

[9]   Pestka, J.: Kathodischer Korrosionsschutz von 110 kV-Gasaußendruckkabeln.
       Rohre, Rohrleitungsbau, Rohrleitungstransport (3R) International Vol. 22 (1983)
       pp. 228–231

[10]   Kaminaga, K. et al.: Development of 500 kV XLPE Cables and Accessories for
       Long-Distance Underground Transmission Lines – Part V: Long-Term Perfor-
       mance for 500 kV XLPE Cables and Joints. IEEE-PES Winter Meeting Baltimore
       MD, January 1996

[11]   Goehlich, L.; Kammel, G.; v. Olshausen, R.: Leitschichten als bedeutende
       Einflußgröße auf die Betriebseigenschaften von VPE-Kabeln. ETG-Fachtagung
       "Isoliersysteme der elektrischen Energietechnik – Lebensdauer Diagnostik und
       Entwicklungstendenzen", Würzburg 1992, Beitrag 3–7; ETG Specialist Report
       No. 40, pp. 339–349

[12]   Meier, N.: Über das Durchschlagverhalten von mechanisch beanspruchtem Poly-
       ethylen im inhomogenen Wechselfeld. Diss. University of Hanover 1975

[13]   Fischer, P.: Electrical Conduction in Polyolefins. Journ. of Electrostatics 4
       (1977/1978), pp. 149–173

[14]   v. Olshausen, R.: Die elektrische Langzeitfestigkeit von Polyethylen. Erschienen
       in "Kunststoffe in der Kabeltechnik", published by H. J. Mair, 2nd edition, Expert-
       Verlag, Ehningen 1991

[15]   Beyer, M.; Boeck, W.; Möller, K.; Zaengl, W.: Hochspannungstechnik. Theoretische und praktische Grundlagen. Springer-Verlag Berlin Heidelberg New York London Paris Tokyo 1986

[16]   Löffelmacher, G.: Über die physikalisch-chemischen Vorgänge bei der Ausbildung von Entladungskanälen in Polyethylen und Epoxidharz im inhomogenen Wechselfeld. Diss. University of Hanover 1975

[17]   Feichtmayr, F.; Würstlin, F.: Die Spannungsstabilisierung von Hochdruck-Polyethylen. Kunststoffe Vol. 60 (1970) pp. 57–62

[18]   Brancato, E. L.: Insulation ageing: A historical and critical review. IEEE Trans. on Electr. Insulation EI-13 (1978) pp. 308–317

[19]   Bahder, G. et al.: Physical model of electric ageing and breakdown of extruded polymeric insulated power cables. IEEE Trans. on PAS, PAS-101 (1982) pp. 1379–1388

[20]   Dorison, E. et al., high-voltage cross-linked polyethylene insulated cables in the French national grid – experience in the field – potential utilization at higher voltages. Cigré, Paris 1994, Contribution 21–107

[21]   Peschke, E. F.; Schroth, R. G.; v. Olshausen, R.: Erweiterung des Einsatzbereiches von VPE-Kabeln auf 500 kV durch technologische Fortschritte. Elektrizitätswirtschaft, year. 94 (1995) pp. 1818–1823

[22]   v. Olshausen, R.; Peschke, E. F.; Schroth, R. G.: Entwicklung von VPE-Kabeln bis 400 kV: Dimensionierung und Prüfphilosophie. etz year. 111 (1990) pp. 1000–1009

[23]   Sato, T.; Muraki, K.; Sato, N.; Sekii, Y.: Recent technical trends of 500kV XLPE cable. IEE 3rd Int. Conf. on Power Cables & Accessories 10kV~500kV, London, November 1993

[24]   Weißenberg, W.; Küllig, P.; Mosch, W.; Eberhardt, M.: The influence of thermooxidized and deteriorated particles of polyethylene as macroscopic imperfections on the lifetime curve at AC stress in PE/XLPE insulated cables. Cigré Symposium, Vienna 1987, Section 6.2

[25]   Großekatthöfer, H.-D.: Der Trocknungsprozeß von Papierisolationen und der Einfluß von Restfeuchte auf die elektrischen Eigenschaften des imprägnierten Dielektrikums. Diss. University of Hanover 1973

[26]   Peschke, E. F.: Einleiter-Ölkabel und Garnituren für 380 kV. Siemens-Zeitschrift year. 50 (1976) pp 676–684

[27]   Fischer, P.; Röhl, P.: Application of statistical methods to the analysis of electrical breakdown in plastics. Siemens R & D reports Vol. 3 (1974) pp 125–129

[28]   v. Olshausen, R.; Westerholt, H.: Volumeneffekte beim Wechselspannungsdurchschlag von Polyethylen-Modellprüfkörpern. etz archive 3 (1982) pp 23–26

[29]   Nagasaki, S. et. al.: Philosophy of design and experience on high-voltage XLPE cables and accessories in Japan. Cigré Session, Paris 1988, Contribution 21–01

[30]   v. Olshausen, R.; Peschke, E. F.; Schroth, R. G.: Development of EHV XLPE Cables: Dimensioning and Test Philosophy. Cigré Session, Paris 1990, Contribution 21–107

[31] Schmidt, B.: Verminderung der Mantelverluste unbewehrter Energiekabel durch Kreuzen der Mäntel in den Verbindungsmuffen. BBC-Nachrichten, February 1963, pp 73–78

[32] Tsuda, H.; Kawazu, M.; Fukagawa, H.; Imajo, T.: Development and actual use of 275 kV low-loss self-contained oil-filled cables with laminated paper insulation. IEEE Trans. on PAS, PAS-103 (1984), pp 744–753

[33] Brand, U.; Kind, D.: Gas-impregnated plastic foils for high-voltage insulation. Cigré Session, Paris 1972, Contribution 15–02

[34] Couderc, D.; Bui Van, Q.; Vallée, A.; Hata, R.; Murakami, K.; Mitani, M.: Development and testing of a 800 kV PPLP-insulated oil-filled cable and its accessories. Cigré Session, Paris 1996, Contribution 21/22–04

[35] Winkler, F.: Strombelastbarkeit von Starkstromkabeln in Erde bei Berücksichtigung der Bodenaustrocknung und eines Tageslastspieles. etz Report 13, VDE-Verlag Berlin and Offenbach, 1978

[36] Große-Plankermann, G.: Grundlagen der Berechnungsmethoden der erarbeiteten Belastbarkeitstabellen. Energiewirtschaftliche Tagesfragen 27 (1977) pp 121–126

[37] Iwata, Z.; Gregory, B. et al.: Garnituren für Hochspannungskabel mit extrudierter Isolierung. Cigré Working Group 21.06 / etz-Report 29, VDE-Verlag Berlin and Offenbach, 1996

[38] Brakelmann, H.: Belastbarkeitsberechnungen für Kreuzungen von Kabeltrassen. etz-Archiv 6 (1984) S. 9-16

[39] Brakelmann, H.: Kabelbelastbarkeits-Reduktion durch thermisch ungünstige Trassenbereiche. Elektrizitätswirtschaft 83 (1984) pp 63–69

[40] Iwata, Z.; Ichiyanagi, N.: New local cooling method for underground power transmission by heat-pipe. IEEE Trans. on Power, App. & Systems, PAS-99 (1980) pp 1038–1046

[41] Imajo, T.; Fukagawa, H.; Itoh, T.: Thermal interference in underground cable heat dissipation by steam pipes and its prevention. 2nd Int. Conf. on Power Cables and Accessories 10 kV to 180 kV London, Nov. 1986. IEE Conf. Publication No. 270, pp 222–226

[42] Kiwit, W.; Wanser, G.; Laarmann, H.: Hochspannungs- und Hochleistungskabel. Verlags- u. Wirtschaftsges. d. Elektrizitätswerke m.b.H. – VWEW – Frankfurt/M 1985

[43] Morello, A.; Hosokawa, K.; Cookson, A. H.; Silver, D. A.: Supertension cables for transmission of large power – Part A: Present state-of-the-art. Cigré-Symposium 06–85, Brussels 1985, Contribution 230–04

[44] Hauge, O. et al.: The Skagerrak HVDC cables. Cigré Session, Paris 1978, Contribution 21–05

[45] Carcano, C. et al.: Qualification test program for the 400 kV HVDC deep water interconnection between Italy and Greece. Cigré Session, Paris 1996, Contribution 21–304

[46] Sugata, T. et al.: Development of 500 kV DC PPLP-insulated oil-filled submarine cable. Cigré Session, Paris 1996, Contribution 21–302

[47]   Last, F. H. et al.: The underground HVDC link for the transmission of bulk power from the Thames estuary to the centre of London. IEEE Trans. on Power, App. & Systems, PAS-90 (1971) pp 1893–1901

[48]   Winkler, F.: Kabel für HGÜ. Contribution zum HGÜ-Kolloquium Heidelberg, 21./22. October 1964

[49]   Weedy, B. M.; Chu, D.: HVDC extruded cables – parameters for determination of stress. IEEE Trans. on Power, App. & Systems, PAS-103 (1984) pp 662–667

[50]   v. Olshausen, R.; Sachs, G.: AC loss and DC conduction mechanisms in polyethylene under high electric fields. IEE Proc. Vol. 128 Pt. A (1981) pp 183–192

[51]   Boone, W.; Kalkner, W.; Weck, K.-H.: Aufgaben und Möglichkeiten der Vor-Ort-Prüfung von Mittelspannungs-Kabelanlagen. Erschienen bei Krefter (Herausgeber): "Prüfungen zur Beurteilung von Kabelanlagen in Mittelspannungsnetzen", VWEW-Verlag Frankfurt/Main 1991, pp 89–114

[52]   Dehoust, O.: Über den elektrischen Leitungsmechanismus in Polyethylen. Kolloid-Zeitschr. u. Zeitschr. f. Polymere 235 (1969) pp 1271–1280

[53]   Röhl, P.; Fischer, P.: Zur elektrischen Gleichstromleitfähigkeit von Polyethylen. Kolloid-Zeitschr. u. Zeitschr. f. Polymere 251 (1973) pp 947–950

[54]   Adamec, V.; Calderwood, J. H.: Electrical conduction in dielectrics at high fields. Journal Physics D: Applied Physics 8 (1975) pp 551–559

[55]   Kelk, E.; Counsell, J. A.H.; Edwards, D. R.: Oil impregnated paper as insulation for high-voltage DC cables. IEE Conf. Publication 22 part 1 (1966) pp 334–337

[56]   Terashima, K. et al.: Research and Development of ± 250 kV DC XLPE Cables. ETG/IEEE PES Conference Berlin, July 1997, Contribution PE-597-PWRD-0-04–1997

[57]   Berger, N.; Randoux, M.; Ottmann, G.; Vuarchex, P.: Review on Insulating Liquids Part 1 – Insulating Liquids: Description and Technical Aspects. Electra No. 171 (April 1997)pp 33–43

[58]   Link, H.: Das Temperaturverhalten der elektrischen Eigenschaften von Isolierölen, insbesondere von Haftmassen und ihren Einzelkomponenten. Diss. TH Braunschweig 1966

[59]   Beyer, M.: Die Vakuumtrocknung von Isolierflüssigkeiten und Papierisolationen und ihr Einfluß auf deren elektrische Eigenschaften. Druckschrift des VDI-Bildungswerkes BW 2192 (1971)

[60]   Jacobsen, P.: Einflußgrößen auf den Trocknungsprozeß von papierisolierten Hochspannungskabeln. Diss. University of Hanover 1977

[61]   Bitsch, R.: Gase und Wasserdampf im Isolieröl und ihr Einfluß auf seine elektrische Festigkeit im inhomogenen Wechselfeld. Diss. University of Hannover 1972

[62]   Holle, K.-H.: Über die elektrischen Eigenschaften von Isolierölen, insbesondere über den Einfluß von Wasser auf deren Temperaturverhalten. Diss. TU Braunschweig 1967

[63]   Bessei, H.; v. Olshausen, R.: Water ingress and moisture effects on MIND cable

dielectrics with respect to the application of polymeric accessories. 2nd IEE Int. Conf. on Power Cables and Accessories 10 kV to 180 kV, London 1986; IEE Conf. Publ. No. 270, pp 119–123

[64]   Krupski, J.; Linke, G.; Peschke, E.: Hochdruck-Ölkabel im Stahlrohr für 380 kV. Siemens-Zeitschrift year 48 (1974) pp 568–575

[65]   Wieland, A.: Durchschagverhalten von SF_6 und SF_6-Gas-Gemischen bis zu hohen Gasdrücken. Diss. TH Darmstadt 1978

[66]   Sünderhauf, H.; Zagorni, V.: Gasinnendruckkabel für Hochspannungsübertragungen. Siemens-Zeitschrift year 41 (1967)pp 38–44

[67]   Peschke, E. F.: Hochleistungsübertragung mit Kabeln. Siemens R & D reports Vol. 2 (1973) pp 46–57

[68]   Bahder, G. et al.: Development of ±400 kV/±600 kV high and medium-pressure oil-filled paper-insulated DC power cable system. IEEE Trans. on Power, App. & Systems, PAS-97 (1978) pp 2045–2056

[69]   Fasching, G.: Werkstoffe für die Elektrotechnik – Mikrophysik, Struktur, Eigenschaften. Springer-Verlag Wien, New York, 1984, pp 225 ff

[70]   Vieweg, R. (Herausg.): Kunststoff-Handbuch: Herstellung, Eigenschaften, Verarbeitung und Anwendung, Band IV Polyolefine. Hanser-Verlag, Munich 1969

[71]   Wagner, H.; Wartusch, J.: About the significance of peroxide decomposition products in XLPE cable insulations. IEEE Trans. on Electrical Insulations EI-12 (1977) pp 395–401

[72]   Goehlich, L.; Kammel, G.; v. Olshausen, R.: Eigenschaften von VPE-Mittelspannungskabeln mit unterschiedlichen Leitschichten. etz Vol. 114 (1993) pp 1184–1188

[73]   Farkas, A. A.; Nilsson, U. H.; Åkermark, G.: High performance semiconductive compounds: Testing, production and experience. Jicable-91 Versailles, June 1991, Contribution A 1.2

[74]   Burns Jr., N. M.: Performance of supersmooth, extra clean semiconductive shields in XLPE insulated power cables. IEEE Int. Symp. on El. Insul., Toronto June 3–6, 1990, Conference Folder pp 272–276

[75]   Schroth, R. G.: discussion contribution to Cigré conference, Paris 1988, group 21, preferred subject No. 1. Proc. of the 1988 Session Cigré Group 21, pp 2-4

[76]   Andreß, B.; Fischer, P.; Röhl, P.: Bestimmung der elektrischen Festigkeit von Kunststoffen. ETZ-A 94 (1973) pp 553–556

[77]   Peschke, E. F.; Schroth, R. G.; v. Olshausen, R.: Extension of XLPE Cables to 500 kV Based on Progress in Technology. Revue de L'Électricité et de L'Électronique, Numéro spécial – Août 1996, pp 19–25

[78]   Schädlich, H.; Klaß, J.: Investigation of XLPE insulations after high stress ageing. Revue de L'Électricité et de L'Électronique, Numéro spécial – Août 1996, pp 59–65

[79]   Kaminaga, K. et al.: Long term test of 500kV XLPE cables and accessories. Cigré Session, Paris 1996, Contribution 21–202

[80] Argaut, P. et al.: Studies and development in France of 400 kV cross-linked poly-ethylene cable systems. Cigré Session, Paris 1996, Contribution 21–203

[81] v. Olshausen, R.; Peschke, E. F.: EHV XLPE cable systems: Reliability by devel-opment testing and quality assurance. Int. Seminar on Cables, Conductors & Winding Wires, Neu Delhi, Nov. 1991

[82] Schroth, R. G.; Kalkner, W.; Fredrich, D.: Test methods for evaluating the water tree ageing behaviour of extruded cable insulations. Cigré Paris 1990, Contribu-tion 15/21–01

[83] Meurer, D.; Kaubisch, D.; Gölz, W.: Zum Langzeitverhalten von Isoliersystemen für VPE-isolierte Mittelspannungskabel. Elektrizitätswirtschaft year 89 (1990), pp 1486–1494

[84] Huth, O.; Maier, R.; Weißenberg, W.: Cleanliness – a Top Priority. Siemens EV-Report Vol. VII, Issue 1, March 1996, pp 21–23

[85] Maier, R.: Fertigung von VPE-Hoch- und Höchstspannungskabeln unter Rein-raumbedingungen. Elektrie, Berlin 51 (1997) pp 62–64

[86] Tanimoto, H.: Manufacture of polyethelene compounds for ultra high-voltage cables. Jap. Patent Nr. 05096534 A2 930420, Showa Electric Wire & Cable Co., Japan, 1991

[87] Heckmann, W.; Schlag, J.: Mikroskopische Untersuchungen an dampfvernetzten Hochspannungsisolierungen aus Polyethylen. Kunststoffe Vol. 72 (1982) pp 96–101

[88] Kalkner, W.; Müller, U.; Peschke, E. F.; Henkel, H.-J.; v. Olshausen, R.: Water treeing in PE- und VPE-isolierten Mittel- und Hochspannungskabeln. Elektrizi-tätswirtschaft Vol. 81 (1982), pp 911–922

[89] Selle, F.: Durchschlagverhalten von Kunststoffisolierungen bei Überlagerung von Gleich- und Stoßspannung. Diss. University of Hannover 1987

[90] Fischer, P.; Peschke, E. F.: Einsatz von Polyethylen und vernetztem Polyethylen als Isolierwerkstoff für Hochspannungskabel. Siemens R & D reports Vol. 10 (1981), pp 197–204

[91] Targiel, G.: Doppelrotationstechnik zur Fertigung von MV- bis EHV-Adern auf CCV-Anlagen. Draht Vol. 46 (1995) pp 634–637

[92] N. N.: Vorrichtung zum SZ-Verseilen elektrischer Kabel. Patent P 24 54 777.1–34 dated 28. 4. 1977 for D. Vogelsberg, Siemens AG, Berlin and Munich

[93] Arkell, C. A.; Ball, E. H.; Barton, A. H.; Beale, H. K.; Williams, D. E.: The design and installation of cable systems with separate pipe water cooling. Cigré Session, Paris 1978, Contribution 21–01

[94] Arkell, C. A.; Bazzi, G.; Ernst, G.; Schuppe, W. D.; Traunsteiner, W.: First 380 kV bulk power transmission systems with lateral pipe external cable cooling in Austria. Cigré Session, Paris 1980, Contribution 21–09

[95] Krupski, J.; Linke, G.; Peschke, E.: Zwangskühlung von Einleiter-Ölkabeln bis 380 kV. Elektrizitätswirtschaft Vol. 74 (1975), pp 934–941

[96]   Müller, U.; Peschke, E. F.; Hahn, W.: The first 380 kV cable bulk power transmission in Germany. Cigré Session, Paris 1976, Contribution 21–08

[97]   Franke, P.; Hierath, E.; Schilling, K.: Mit Höchstspannung an die Spree – Berliner Umspannwerk Reuter erhält von Westen her Anschluß an das 380 kV-Netz. Siemens EV Report 4/94, pp 17–21

[98]   Brotherton, W.: Field trials of 400 kV oil-cooled cables. Proc. IEE 124 (1977) pp 326–333

[99]   N. N.: Advantages of 400 kV internally oil-cooled cables. Electric. Times 158 (1970) pp 59–60

[100] N. N.: Das neue 110 kV-Hochleistungskabel mit innerer Wasserkühlung – ein Großversuch in Berlin. Elektrizitätswirtschaft 78 (1979), pp 848–851

[101] Friedrich, J.: Endverschlüsse für leitergekühlte Hochleistungskabel – Wirkungsweise und Versuchserfahrungen. etz-a 98 (1977) pp 485–487

[102] Barnes, C. C.; Miranda, F. J.; Hollingsworth, P. M.: Power Ratings of 275 kV and 400 kV Cables in the British Transmission System. Cigré Paris 1966, Contribution 207

[103] Arkell, C. A. et al.: The development and application of forced cooling techniques to EHV cable systems in the U. K. Cigré Session, Paris 1976, Contribution 21–02

[104] Watanabe, T.; Tsuda, H.: The technical development and practical use of bulk power underground transmission cable systems in Japan. Cigré Session, Paris 1978, Contribution 21–03

[105] Helling, K.; Henningsen C. G.; Polster, K.; Schroth, R. G.: Power supply for the city of Berlin: Are 380 kV XLPE cables a safe alternative to long-time proved LPOF-cables? Cigré Session, Paris 1994, Contribution 21–106

[106] Aucourt, C. et al.: Gas insulated cables: from the state of the art to feasibility for 400 kV transmission lines. Jicable 1995, Contribution A.5.4

[107] Thuries, E. et al.: 420 kV three-phase compressed nitrogen insulated cable. Jicable 1995, Contribution A.5.2

[108] N. N.: SF_6-isolierte Rohrleiter für 420/525 kV. Product sheet E 121/1475, Siemens AG

[109] Deschamps, L. et al.: Compressed gas insulated cables in use internationally. Electra, 1984, No.94, pp 55–70

[110] Kuroyanagi, Y. et al.: Construction of 8000 A class 275 kV gas insulated transmission line. IEEE Transmission on Power Delivery, Jan. 1990, No.1, pp 14–20

[111] Kaminaga, K. et al.: Development of compact 500 kV 8000 A gas insulated transmission line. IEEE Transmission on Power Delivery, Oct. 1987, No.4, pp 961–968

[112] Artbauer, J.: Hochspannungskabel mit SF_6-Isolierung. Elektrizitätswirtschaft 81 (1982), pp 21–26

[113] Miller, D. B. et al.: Flexible gas insulated cable for 220 kV, 345 kV and 500 kV. IEEE Trans. on Power, App. & Systems, PAS-103 (1984) pp 2480–2485

[114] Spencer, E. M. et al.: R&D of a flexible 345 kV compressed gas insulated transmission cable. Cigré Session, Paris 1980, Contribution 21–02

[115] Cousin, V, Koch, H.: From gas-insulated switchgear to cross country cables. Jicable 1995, Contribution A 5.1

[116] Koch, H. et al:. Pipeline für den Strom. Siemens EV-Report 3/95, pp 8–9

[117] Neumüller, H. W.: Hochtemperatur-Supraleitung für die Energietechnik. 16. Hochschultage Energie, Sept. 1995, RWE Energie AG, Essen; Conference report pp 51–75

[118] Birnbreier, H. et al.: Energieübertragung mit Kryokabeln. Jülich atomic energt research plant, study Jül-938-TP, March 1973

[119] Erb, J. et al.: Comparison of advanced high power underground cable design. Gesellschaft für Kernforschung, Karlsruhe, study KFK 2207, September 1975

[120] Bogner, G. et al.: Development of a superconducting high power AC cable. Siemens R & D reports Vol. 8 (1979), Part 1–3, pp 1–22

[121] Forsyth, E. B.; Thomas, R. A.: Performance summary of the Brookhaven superconducting power transmission system. Cryogenics, 26 (1986), pp 599–614

[122] N. N.: Elektrische Hochleistungsübertragung und -verteilung in Verdichtungsräumen. FGH Mannheim, study 1977

[123] Bogner, G.; Neumüller, H. W.: Hochtemperatur-Supraleiter für die Energietechnik. Siemens-Zeitschrift, Special FuE, Herbst 1995, pp 32–35

[124] Bogner, G. et al: The advance of high T_C superconductors and their impact on the application of superconductivity in the electric industry. Cigré WG 11–05, Electra No. 114 (1987) pp 96–107

[125] Oswald, B. R.: HTSL-Anwendungsmöglichkeiten in der Energietechnik. etz year 118 (1997) H. 3, pp 52–53

[126] Henningsen, C. et al.: Übertragungsleitungen. ETG-Fachbericht Nr. 35 (1991) pp 59–71

[127] Hardtke, R.; Henningsen, C.-G.; Polster, K.: 380 kV diagonal connection through the load centres in Berlin. Tech. report, Bewag Berlin, 1997

[128] Uchida, K. et al.: Study on detection for the defects of XLPE cable lines. IEEE Trans. on Power Delivery, Vol. 11 (1996) pp 663–668

[129] Helling, K. et al.: Pre-qualification test of 400 kV XLPE cable systems. Jicable 1995, Contribution A.1.3

[130] Lesch, G.: Lehrbuch der Hochspannungstechnik. Springer-Verlag Berlin / Göttingen / Heidelberg 1959

[131] Oesterheld, J.; v. Olshausen, R.; Pöhler, pp: Optimized design of accessories for 245 kV and 420 kV XLPE cables. Cigré Session, Paris 1992, Contribution 21–202

[132] Kunze, D.: Montagefreundlich und umweltverträglich: Eine neue Füllmasse für Garnituren von VPE-Hochspannungskabeln. ETG-Tage Mannheim, 19.–21.10.1993, Contribution to theme group "Werkstoffe in der Energietechnik" (tech. area 9)

[133] Favrie, E.; Pays, M.: Present practices for underground transmission cables in France. I.E.E.E. T and D Conf., Chicago, 10–15 April 1994; Panel Session on Worldwide Underground Transmission Practices.

[134] Onodi, T.: Prefabricated accessories for high-voltage cables – design, dimensioning and field experience. Jicable 1995, Contribution A.4.2.

[135] Vasseur, E.; Chatterjee, pp: Development of HV and EHV single piece pre-moulded joint. Jicable 1995, Contribution A.4.4.

[136] Andersen, P. et al.: Development of a 420 kV XLPE cable system for the metropolitan power project in Copenhagen. Cigré Session, Paris 1996, Contribution 21–201

[137] Peschke, E. F.; Kunze, D.; Schoth, R. G.; Weißenberg, W.: A new generation of joints for XLPE-insulated extra-high-voltage cables. Cigré Session, Paris 1996, Contribution 21–204

[138] Kunze, D.; Oesterheld, J.: Silikonkautschuk – ein wichtiger Werkstoff in der Garniturentechnik für Hoch- und Höchstspannungs-VPE-Kabel. Elektrizitätswirtschaft 94 (1995) pp 1859–1864

[139] Fricke, W.; Kunze, D.; Scharschmidt, J.; Weißenberg, W.: Endverschlüsse im Baukastensystem. Siemens EV Report, year 8 (1997) Ausgabe 2, pp 24–25

[140] Kikuchi, K. et al.: Recent technical progress in accessories for extra high-voltage XLPE cables in Japan. Cigré Session, Paris 1992, Contribution 21–203

[141] Jahnke, B.; Kohlmeyer, A.: Erste 220 kV-VPE-Kabelanlage mit vorgefertigten Verbindungsmuffen in Deutschland. Elektrizitätswirtschaft 94 (1995) pp 1824–1830

[142] Argaut, P.; Favrie, E.: Recent developments in 400 and 500 kV XLPE cables. Jicable 1995, Contribution A.1.1.

[143] Rosevear, R. D.; Williams, G.; Parmigiani, B.: high-voltage XLPE cable and accessories. IEE 2nd Int. Conf. on Power Cables and Accessories 10 kV to 180 kV, London 1986; IEE Conf. Publ. No. 270, pp 232–237

[144] Imajo, N. et al.: The Reduced Insulation Thickness 275 kV XLPE Cable and Its Extrusion Type Molded Joint. Fujikura Technical Review 1993, pp 16–22

[145] Kubota, T. et al.: Development of 500 kV XLPE Cables and Accessories for Long-Distance Underground Transmission Lines – Part II: Jointing Techniques. IEEE 1994 PES Winter Meeting, 94WM040–6PWRD

[146] Takeda, N. et al.: Development of 500 kV XLPE Cables and Accessories for Long-Distance Underground Transmission Lines – Part IV: Electrical Properties of 500 kV Extrusion Molded Joints. IEEE Trans. on Power Delivery Vol. 11 (1996), pp 635–643

[147] Fuhrmann, B.: 110 kV-Übergangsmuffe: Ein systemgerechtes Bindeglied zwischen konventioneller Ölkabel- und VPE-Kabeltechnik. Elektrizitätswirtschaft year. 91 (1992) pp 1743–1744

[148] N.N.: Kabelhandbuch / 4. Auflage. Herausgegeben von der Vereinigung Deutscher Elektrizitätswerke – VDEW e.V. Verlags- und Wirtschaftsgesellschaft der Elektrizitätswerke – VWEW, Frankfurt (Main) 1986

[149] Angenend, M.; Hahne, G.: Kabelmontage. ETG specialist conference "110 kV-Kabelanlagen in der Städtischen Stromversorgung", Essen, 25.03.1993

[150] Jahnke, B.: Kabelbauarten. ETG specialist conference "110 kV-Kabelanlagen in der Städtischen Stromversorgung", Essen, 25.03.1993

[151] v. Olshausen, R.: Garnituren und Anlagentechnik. ETG specialist conference "110 kV-Kabelanlagen in der Städtischen Stromversorgung", Essen, 25.03.1993

[152] Brandes, W.: Planungsprämissen 110 kV-Kabelnetze. ETG specialist conference "110 kV-Kabelanlagen in der Städtischen Stromversorgung", Essen, 25.03.1993

[153] Korobbokova, V.P. et al.: Influence of the electric field in 500 and 750 kV switchyards on maintenance staff and means for its protection. Cigré Session, Paris 1972, Contribution 23–06

[154] Schneider, K.-H. et al.: Displacement currents to the human body caused by the dielectric field under overhead lines. Cigré Session, Paris 1974, Contribution 36–04

[155] Schmidt, H.-G.; Tolazzi, H.: Art und Umfang der Gefährdungsmöglichkeiten der Arbeitnehmer an Arbeitsplätzen in elektrischen Hochspannungsanlagen durch die Einwirkung elektrischer und elektromagnetischer Felder. Report to Fed. German Ministry of Employment and Sozial Welfare 1974

[156] Phillips, R.D.; Kaune, W.T.: Biological effects of static and low-frequency electromagnetic fields: an overview of United States literature. EPRI EA-490-SR, Special Report, July 1977

[157] Kühne, B.: Methoden zur Untersuchung des Einflusses hoher elektrischer 50-Hz-Felder auf den menschlichen Organismus. Diss. University of Hannover 1979

[158] Haubrich, H.-J.: Das Magnetfeld im Nahbereich von Drehstrom-Freileitungen. Elektrizitätswirtschaft 73 (1974) pp 511–517

[159] Endersby, T.M.; Galloway, pp J.; Gregory, B.; Mohan, N.C.: Environmental compatibility of supertension cables. IEE 3rd Int. Conf. on Power Cables & Accessories 10kV~500kV, London November 1993, Session 4A

[160] Vérité, J.C. et al.: Magnetic fields in HV cable systems 1: systems without ferromagnetic component. Bericht der Cigré Joint Task Force 36.01/21, June 1996

[161] Harjes, B.; Peier, D.; Senftleben, H.: Energietechnische LWL-Anwendungen. etz year 113 (1992) pp 984–990

[162] Brakelmann, H.: Belastungsmonitor für Energiekabeltrassen. etz Vol. 1–2/1997 pp 30–34

[163] Tozaki, T. et al.: Raman backscattering characteristics of the optical fiber and distributed temperature sensor. Fujikura Technical Review 1990

[164] Glaese, U.; Goehlich, L.: Überwachung von Hochspannungskabelanlagen – Methoden und Kundennutzen. Elektrizitätswirtschaft 94 (1995) pp 992–1000

[165] Glaese, U.: Entwicklung eines automatisierten Monitoring-Systems für Hochspannungskabel. Dissertaion Universität Hannover / Fortschr.-Ber. VDI Series 21 No. 212, VDI-Verlag Düsseldorf 1996

[166] Ogawa, K. et al.: Humidity-sensing effects of optical fibres with microporous SiO$_2$-cladding. Electronics Letters Vol. 24 (1988) No. 1, pp 42–43

[167] Aihara, M. et al.: Insulation monitoring system for XLPE cable containing water sensor and optical fiber. Proc. of 3rd Int. Conf. on Properties and Appl. of Dielectric Materials, Tokyo, 8.–12. July 1991

[168] Pommerenke, D.; Krage, I.; Kalkner, W.; Lemke, E.; Schmiegel, P.: On-site PD measurement on high-voltage cable accessories using integrated sensors. 9th Int. Symp. on HV Engineering, Graz 1995

[169] Miyazaki, T. et al.: Partial discharge measurement system for field testing of XLPE cable. Sumitomo El. Technical Rev. No. 39 (1995) pp 38–42

[170] Krage, I.; Strehl, T.; Kalkner, W.: Measurement and location of partial discharges during on-site testing of XLPE cables with impulse voltages. Jicable 1995, Contribution D.3.6

[171] Schichler, U.: Erfassung von Teilentladungen an polymerisolierten Kabeln bei der Vor-Ort-Prüfung und im Netzbetrieb. Diss. Univ. Hanover 1996

[172] Jacobsen, P.: SF$_6$-Sensor in Schaltereinführungsendverschlüssen von Niederdruck-ölkabeln. etz (1995) Heft 8, pp 30–32

[173] N. N.: Richtlinien zum umfassenden Qualitätsmanagement. European Foundation for Quality Management, Brussels 1997

[174] N. N.: Unternehmensqualität: Grundsätze des Siemens-Qualitätsmanagements. Siemens AG, Order No. A 12013-A1, October 1996

[175] Schroth, R. (Convenor) et al.: Recommendations for electrical tests pre-qualification and development on extruded cables and accessories at voltages >150 (170) kV and ≤400 (420) kV. Electra No. 151 (Dec. 1993) pp 15–19

[176] Kubota, T. et al.: Development of 500 kV XLPE Cables and Accessories for Long-Distance Underground Transmission Lines – Part I: Insulation design of cables. IEEE 1994 PES Winter Meeting, 94WM097–6PWRD

[177] Klockhaus, H.; Merschel, F.: Langzeitprüfungen an VPE-isolierten MS-Kabeln – Typprüfung und fertigungsbegleitende Prüfungen. Elektrizitätswirtschaft, year 94 (1995) pp 1808–1816

[178] Schroth, R. (Convenor) et al.: Recommendations for electrical tests type, sample and routine on extruded cables and accessories at voltages >150 (170) kV and ≤400 (420) kV. Electra No. 151 (Dec. 1993) pp 21–29

[179] N. N.: Endverschlußsysteme für Kabelprüfungen Reihe CTTS. publication D 178.10 from Emil Haefely & Cie AG 3.1990

[180] Schädlich, H.: Technische Kennwerte, Normen, Prüfungen von 110 kV-Kabeln. ETG specialist conference "110 kV-Kabelanlagen in der Städtischen Stromversorgung", Essen, 25.03.1993

[181] v. Olshausen, R.: Aktuelle Probleme und Entwicklungstendenzen bei Kabeln. ETG specialist conference "Isoliersysteme der elektrischen Energietechnik – Lebensdauer Diagnostik und Entwicklungstendenzen", Würzburg 1992; ETG-Fachbericht No. 40, pp 277–289

[182] Peschke, E. F.: Spannungsprüfungen an Kabelanlagen. Elektrizitätswirtschaft, year.85 (1986) pp 691–692

[183] Kobayashi, S. et al.: Study on detection for the defects of XLPE cable links. Jicable 1995, Contribution A.6.3.

[184] Parpal, Jl.; Guddemi, C.; Amyot, N.; David, E.: DC testing of XLPE insulation. Jicable 1995, Contribution A.6.2.

[185] Grönefeld, P.; v. Olshausen, R.; Selle, F.: Fehlererkennung und Isolations-gefährdung bei der Prüfung water tree-haltiger Kunststoffkabel mit Spannungen unterschiedlicher Form. Elektrizitätswirtschaft, year. 84 (1985) pp 501–506

[186] Bach, R.; Kalkner, W.: Vergleichende Untersuchungvon alternativen Span-nungsarten für die Vor-Ort-Prüfung von Mittelspannungskabeln. Publication from Krefter (publisher): "Prüfungen zur Beurteilung von Kabelanlagen in Mittelspan-nungsnetzen", VWEW-Verlag Frankfurt/Main 1991

[187] Aucourt, C. et al.: Recommendations for a new after laying test method for high-voltage extruded cable systems. Cigré Session, Paris 1990, Contribution 21–105

[188] Weißenberg, W.; Goehlich, L.; Scharschmidt, J.: Inbetriebnahmeprüfungen von VPE-isolierten Hochspannungskabelanlagen mit Wechselspannung. Elektrizitäts-wirtschaft, year 96 (1997) pp 400–409

[189] Hauschild, W.; Schufft, W.; Spiegelberg, J.: Alternating voltage on-site testing of XLPE cables: The parameter selection of frequency-tuned resonant test systems. 10th Int. Symp. on high-voltage Engineering, Montreal, August 1997

[190] Gockenbach, E.; Schiller, G.: The breakdown behaviour of XLPE samples at volt-ages of diferent shapes. 9th Int. Symp. on high-voltage Engineering, Graz 1995

[191] Steckel, R.-D.; Stubbe, R.: Perspektiven für eine zukünftige Normung von Verteilungskabeln. VDEW-Kabeltagung Nürnberg, 13./14.10.1997

[192] Stubbe, R.: Vorschriften und Normung. Erscheint in: Kabelhandbuch / 5. Auflage. Herausgegeben von der Vereinigung Deutscher Elektrizitätswerke – VDEW e.V. Verlags- und Wirtschaftsgesellschaft der Elektrizitätswerke – VWEW, Frankfurt (Main) 1997

[193] N. N.: Single-core cables with polymeric insulation for rated voltages above 150 kV ($U_m = 170$ kV) up to 500 kV ($U_m = 525$ kV). French Standard C 33–253

[194] Peschke, E. F.: Neue Werkstoffe und gesteigertes Umweltbewußtsein fordern neue Wege in der Hochspannungskabeltechnik. Siemens EV Report; 3rd year. (1992) pp 22–25

[195] Pays, M.: 1995-Progress-Report of SC 21 (high-voltage insulated cables). Electra, No. 166 (1996) pp 69–75

[196] Pays, M.: General Report for Group 21. Cigré Session, Paris 1996; Electra, No.170, (1997) pp 59–73

[197] N. N.: Discussion Proceedings of Group 21. Cigré Session, Paris 1996, Vol. I, Gr. 21

287

[198] Pays, M.: 1996-Progress-Report of SC 21. Electra, No.172 (1997) pp 73–79

[199] Wiznerowicz, F.: Kabeltechnik auf der Internationalen Hochspannungskonferenz (Cigré) 1996 in Paris. Elektrizitätswirtschaft, year. 95 (1996) pp 1789–1793

[200] Orton, H. E.; Samm, R.: Worldwide Undergrund Transmission Cable Practices. IEEE Transaction on Power Delivery Vol.12 (1997) pp 533–541

[201] Peschke, E. F. u.a.: Vernetztes Polyethylen – der Kabelisolierwerkstoff für 1 kV bis 550 kV. Siemens-Zeitschrift Special, Autumn 1996, pp 23–26

[202] N. N.: Épure 1995. Electricité de France, DER, Clamart No. 48, Oct. 1995, pp 13–37 & pp 55–79

[203] Hahne, G.; Waschk, V.: 110 kV-Stadtkabel zum Retrofitting von Rohrkabeln. Elektrizitätswirtschaft, year 95 (1996) pp 1770–1774

Index